192
Advances in Anatomy Embryology and Cell Biology

Wijnand F. R. M. Koch · Enrico Marani

Early Development of the Human Pelvic Diaphragm

With 64 Figures and 2 Tables

Springer

Wijnand F.R.M. Koch, Dr.
Medisch Centrum Alkmaar
Polikliniek Urologie
Wilhelminalaan 12
1815 JD Alkmaar
The Netherlands
e-mail: w.f.r.m.koch@mca.nl

Enrico Marani, Prof. Dr.
Department of Biomedical Signal and Systems
University of Twente
PO Box 217
7500 AE Enschede
The Netherlands
e-mail: e.marani@utwente.nl

ISSN 0301-5556
ISBN 978-3-540-68006-3 Springer Berlin Heidelberg New York

Springer is a part of Springer Science+Business Media
springer.com

Editor: Simon Rallison, Heidelberg
Desk editor: Anne Clauss, Heidelberg
Production editor: Nadja Kroke, Leipzig
Cover design: WMX Design Heidelberg
Typesetting: LE-TEX Jelonek, Schmidt & Vöckler GbR, Leipzig
Printed on acid-free paper SPIN 11890041 27/3150/YL – 5 4 3 2 1 0

List of Contents

Abbreviations

ATLA	Arcus tendineus levatoris ani
CRL	Crown–rump length
HE	Hematoxylin–eosin
POP	Pelvic organ prolapse

1 Introduction and Aim

A sound and detailed knowledge of the anatomy of the pelvic floor is of the utmost importance to gynecologists, obstetricians, surgeons, and urologists, since they all share the same responsibility in treating patients with different pathological conditions caused by pelvic floor dysfunction (Lawson 1974; Mostwin 1991; Klutke 1995; Haderer 2002). The most common clinical expressions of pelvic floor dysfunction are urinary incontinence, anal incontinence, and pelvic organ prolapse (POP). Most often these clinical expressions are found in women, and they are briefly discussed below based on the outline presented in the Third International Consultation on Incontinence, a joint effort of the International Continence Society (ICS) and the World Health Organization (WHO) (Abrams 2005).

The prevalence of severe or significant urinary incontinence ranges between 3% and 17%, with stress urinary incontinence being the most common: approximately 50% of cases. Established potential risk factors are age, childbearing, and obesity. The pelvic floor plays an important role in these risk factors. There is evidence that the pelvic floor structures change with age, giving rise to dysfunction. Pregnancy, and especially vaginal delivery, may result in pelvic floor laxity as a consequence of weakening, stretching, and even laceration of the muscles and connective tissue, or due to damage to pudendal and pelvic nerves. Comparable to pregnancy, obesity causes chronic strain, stretching, and weakening of muscles, nerves, and other structures of the pelvic floor.

Conservative management of stress urinary incontinence focuses primarily on weight loss and physical therapy, both for prevention and treatment. Physical therapy can comprise pelvic floor muscle training, electrical stimulation, magnetic stimulation, and application of weighted vaginal cones. All are based on the principle of improving strength and/or timing of the pelvic floor muscle contraction. Surgical treatment aims to improve support of urinary sphincter and bladder neck, and is performed in the region of the urogenital diaphragm.

Anal incontinence is the involuntary loss of feces (solid or liquid) and flatus. Its prevalence increases with age, varying from 1.5% in children to more than 50% in nursing home residents, and it is almost as common in men as in women. Age is an important risk factor for anal incontinence, because of co-morbidity from surgery, progressive neurological disease, and stroke. Conflicting data exist for the role of childbirth and mode of delivery as risk factors. Other risk factors are iatrogenic factors: episiotomy or anal sphincter myotomy, for the treatment of anal fissures and hemorrhoids, respectively.

Conservative treatments focus on patient education, dietary counseling, management of constipation, and pelvic floor muscle exercises with or without biofeedback. Nonconservative treatment can consist of either pharmacotherapy or surgery. When there is a sphincter defect, sphincteroplasty is performed. In the absence of sphincter defects, sacral nerve stimulation, dynamic graciloplasty, artificial sphincter implantation, or even a diversion can be considered. In contrast to urinary incontinence, the role of the pelvic floor appears to be limited in the management of anal incontinence, and treatment is mainly focused on the anal sphincter complex.

POP is defined as the downward descent of the pelvic organs, resulting in a protrusion of the vagina and/or the uterine cervix; it does not include rectal prolapse. It is a common condition, affecting 50% of parous women, with a 7% lifetime risk for undergoing surgery for prolapse by age 80. Risk factors for developing POP are not well defined, but they include age, pregnancy, parity, obstetric factors, obesity, race, and gynecologic surgery. Pelvic floor muscle strength progressively decreases with increasing age and parity, and both vaginal childbirth and age have been implicated as major inciting events for pelvic neuropathy, causing POP as a consequence of the resulting pelvic floor dysfunction.

Conservative management of POP is aimed at increasing strength, endurance, and support of the pelvic floor muscles. The role of physical therapy in the treatment of POP, however, remains unestablished. Nonconservative treatments focus on the actual reduction of prolapsing tissues, by resection or by repositioning the tissue to its original position (e.g., sacrocolpopexy).

Judged by the ever-increasing number of papers published in recent years, we believe that more and more attention is given to the physiological aspects of the pelvic floor in maintaining continence and support of pelvic viscera (Barbaric 2001; Olsen 2001; Abrams 2005). Medical specialists appear to progressively realize the enormous advantage of a shared knowledge in their treatment of pelvic floor dysfunction. In this context, it is difficult to understand the striking disregard of medical science for, and unacquaintedness with, embryological anatomy and development of the pelvic floor. Up until today very little is published on this subject (Bitoh et al. 2002; Fritsch 1992, 1993, 1994; Fritsch and Kühnel 1992; Fritsch and Fröhlich 1994; van der Putte 2005).

Our aim is to describe the embryology of the human pelvic diaphragm muscles. A detailed description of the mesenchyme condensations, and the developing muscles originating from them, together with the circumference of their fasciae, will elucidate the differences of opinion on the subdivisions of the pelvic diaphragm muscles. We hope it will also help to improve the rationale of the many therapeutic avenues for pelvic floor dysfunction.

1.1 Review of the Literature

In humans, the functions which the pelvic floor is called upon to perform differ widely from those in animals, in which the long axis of the body is horizontal rather than vertical. In humans, the weight of the abdominal and pelvic viscera is sustained mainly by the floor of the pelvic cavity, and accordingly this structure is specially modified (Thompson 1899; Marani 2002).

In most mammals, the pelvic floor is largely sphincteric in action, to secure continence, as the ventral abdominal wall achieves the supportive function of the (pelvic) viscera. The muscular fibers, which form a distinct layer in the pelvic floor of mammals, are all situated around the rectal and genitourinary canals, and the rest is made up of connective tissue and integuments (Thompson 1899).

In those mammals in which the long axis of the body is absolutely or approximately vertical (e.g., *Homo sapiens*), a great difference in the architecture of the pelvic floor is apparent. The outlet of the pelvic cavity is modified and adapted for supportive maintenance of pelvic and abdominal viscera. In addition to a layer of muscles with mainly sphincteric functions, a second layer has developed into a well-marked diaphragm, which constitutes a muscular and tendinous sheet attached to the inner circumference of the pelvis (Fritsch and Hotzinger 1995; Occelli et al. 2001; Pit et al. 2003). In this pelvic floor two distinct layers or diaphragms can be recognized, with contrasting arrangement and function. The upper or superior layer forms a more or less complete *pelvic diaphragm* (first so named by Meyer in 1861), designed for purposes of support, and the inferior layer forms sphincters around the rectal and genitourinary canal for purposes of maintaining continence. The sphincters around the genitourinary canal are embedded in the urogenital diaphragm, and the rectal sphincters are a separate entity forming the rectal sphincter complex. The two layers are not only different in function but also in morphology. The existence of an embryologic (and consequentially also a neuroanatomic) separation between the two layers of the pelvic floor has been discussed extensively in the medical literature (Marani et al. 1993; Fucini et al. 1999; Marani 2002).

Although the muscles of the pelvic diaphragm are mainly responsible for supporting the viscera, the pelvic bones also have a supportive function (Marani 2002). Davies (1955) has suggested that when humans assumed an upright position, they did so by developing a lordotic curvature of the lumbar spine. When a human stands erect the abdominal axis becomes vertical, but the pelvic axis remains nearly horizontal. This supine posture has been shown to play a role in shaping the human pelvis (Abitbol 1989). Forces on the pelvic floor exerted by gravity and intra-abdominal pressure are guided to the abdominal muscles and pubic bone, and via the ilial bones to the legs, leaving a lesser force to be supported by the pelvic diaphragm (Marani 2002).

1.2 The Upper Layer or Pelvic Diaphragm

In the upper supportive layer, the muscles are derived from the flexors and abductors of the caudal end of the vertebral column; evolutionary, these flexors and abductors have undergone retrogressive changes as the number of caudal vertebrae is reduced in the development to an upright position. Their function has been modified from moving a tail to supporting the viscera (Thompson 1899; Gray and Williams 1995).

The upper supportive layer is invariably known as the pelvic diaphragm and can be divided into two separate layers: the *endopelvic fascia* and the *muscles and fascias of the levatores ani.* Its form is variably described as a basin or a funnel (Waldeyer 1899; Dickinson 1889; Marani 2002), or as a dome (Hjartardottir et al. 1997). The differences can well be explained by the fact that Waldeyer and Dickinson used cadavers for their studies, while nowadays functional studies can be performed by magnetic resonance imaging in living human subjects (Hjartardottir et al. 1997).

1.2.1 (Endo)pelvic Fasciae

The endopelvic or pelvic fascia is a fibromuscular layer consisting of collagen, elastin, and smooth muscle cells, forming a network of sheaths extending from the pelvic floor to the thoracic cavity (De Blok 1982a). One of the greatest sources of confusion has been the attempt to show that it could be traced as one continuous layer over the entire pelvic contents (Barnes 1921). It anchors the postnatal pelvic organs (corpus pelvinum) to the pelvic wall by septa, extending from the pelvic wall to the organs (De Blok 1982b; Marani 2002). It comprises the parietal pelvic fascia and the visceral pelvic fascia.

The *parietal pelvic fascia* is part of the general layer covering the inner aspect of the abdominal and pelvic walls (Gardner et al. 1986), where it forms a part of the pelvic floor and lines the lateral walls of the pelvis (obturator fascia). The parietal pelvic fascia is arranged in two layers: a superior and an inferior fascia. The superior fascia forms the tendinous arch of the pelvic fascia (Pit et al. 2003), of which the most anterior part forms in fact the medial puboprostatic (pubovesical) ligaments (see below).

Posteriorly, this fascia is incomplete: it is absent in front of the sacrum, where a rectosacral fascia is found (Havenga 1998; Marani 2002), thought to anchor the rectum to the sacrum and thus prevent prolapse. Its existence is denied by others (Fritsch and Hotzinger 1995). The part of the parietal fascia covering the obturator internus muscle is called the *obturator fascia*, being attached around the margins of this muscle and covering its pelvic surface. This fascia thus consists of two layers: the superior fascia of the parietal pelvic fascia, and the fascia (or perimysium) of the internal obturator muscle itself. Below the level of the obturator canal, a thickening of the obturator fascia may be present, representing the line of fusion of the obturator fascia with the superior and inferior fascia of the pelvic diaphragm. It extends from the back of the body of the pubis to the ischial spine, giving origin to a part of the levator ani. It is known as the *tendinous arch of the levator ani* (or arcus tendineus levatoris ani). Barnes (1921) stated that in only 40% of cases the part of the levator ani called the iliococcygeal muscle in fact originates from this tendinous arch. In the other 60%, the arcus tendineus levatoris ani was absent and the levator ani took origin from the *arcus tendineus fasciae pelvis* (the white line), a tendinous arch stretching from the posteroinferior border of the pubic bone and symphysis to the ischial spine and sacroiliac joint, forming an Y (Pit et al. 2003). This arch serves to fix the bladder and the prostate to the wall and floor of the pelvis, forming the puboprostatic ligaments and lateral true ligaments of the bladder (Derry 1907). It is in this sense important to keep in mind that the part of the obturator internus muscle below this tendinous arch is in fact extrapelvic.

The *visceral pelvic fascia* is formed by extraperitoneal tissue, variable in structure, serving as a packing for organs and as sheaths for vessels and nerves, connecting them to the parietal pelvic fascia and the inner aspect of the pelvic peritoneum. Some of these sheaths are called ligaments, supporting different endopelvic organs, and are described with the organs they are associated with (De Caro 1998; De Blok 1982a). At the point where pelvic organs pass through the pelvic floor, the parietal and visceral pelvic fascias are continuous.

1.2.2 Musculus Levator Ani

The m. levator ani (first called m. sedem atollens; Vesalius 1555) is a broad muscular sheet of variable thickness attached to the internal surface of the true pelvis, i.e., the back of the body of the pubic bone, to the pelvic fascia on the sidewalls (tendinous arch of levator ani and pelvic fascia) and to the spine of the ischium. Its upper surface is covered by a distinct fascia, its inferior surface by a very thin fascia, better called a perimysium (Barnes 1921).

The difference of opinion concerning the anatomy of the levator ani is well described by Dickinson (1889): ". . . there is no considerable muscle in the body whose form and functions are more difficult to understand than those of the levator ani, and about which such nebulous impressions prevail."

Some authors describe the levator ani as divided into two parts: the pubococcygeus and the iliococcygeus muscle (Elftman 1932; Thompson 1899; Waldeyer 1899). Others include the puborectalis muscle and/or the coccygeus muscle with the levator ani (Mostwin 1991), or typically exclude it (Shafik 1975). Difference of opinion exists about whether the two muscles are attached to or overlapping each other (Knauer 1914; Elftman 1932) or are demonstrably separate (Quain's Anatomy 1899; Lawson 1974a, b; Waldeyer 1899). Lartschneider (in Thompson 1899) spoke of a "portio pubica" and "portio iliaca," indicating that the pubococcygeal and iliococcygeal muscles cannot be separated. Differences can also be noted in the origin of these two muscles: invariably the pubococcygeus arises from some part of the pubic bone, whereas the iliococcygeus originates from the "white line" but can extend in front of the obturator canal to the pubic bone. The separation of pubococcygeus from iliococcygeus is then based on an imaginary line running from the anterior edge of the ischial tuberosity to the junction of the superior pubic ramus with the ilium (Thompson 1899; Holl 1897; Zacharin 1985). Fritsch and Fröhlich (1994) showed that in human development the tendinous arch of the pelvic fascia plays an important role in the differentiation of pubococcygeus and iliococcygeus.

Lawson (1974a, b) divided the levator ani muscles into two different functional groups, pubovisceral and diaphragmatic, based on detailed studies of the pelvic floor in neonates and children. The pubovisceral group is composed of muscles originating from the back of the pubic bone and inserting directly or indirectly into the pelvic viscera (Lawson describes puborectalis, puboanalis, pubovaginalis and puboprostatic muscles). In addition to a supportive function these muscles can also move the viscera upward and forwards. The diaphragmatic, or posterior, group has a simple supporting, diaphragmatic role. These muscles originate from the tendinous arch and ischial spine, forming a continuous sheet, which inserts via three main parts (pubococcygeus, iliococcygeus, and coccygeus) into the anterior sacrococcygeal ligament and the bodies of the lower sacral and coccygeal vertebrae. The most anterior fibers curve back to form an inverted arch with the fibers from the opposite side.

Figure 1 shows an interpretation of the pelvic diaphragm divided into separate muscles, as described above. Others, like Thompson (1899), Holl (1897) and Zacharin (1985), describe the levator ani as a single muscle (Fig. 2).

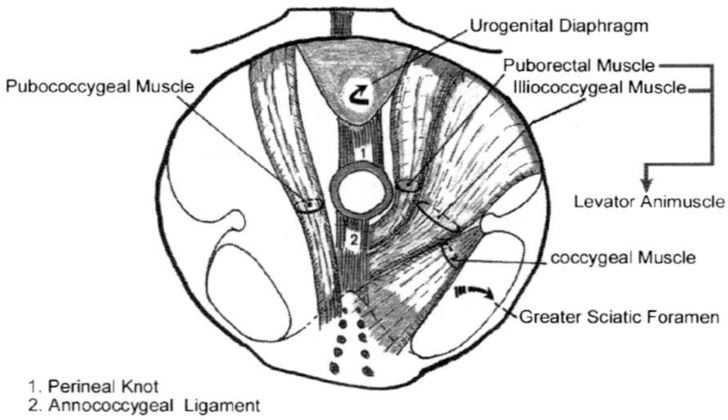

Fig. 1 Muscles of the pelvic diaphragm, as separate muscles. (From Marani 2002, with permission) (full color version of this illustration at the end of the book)

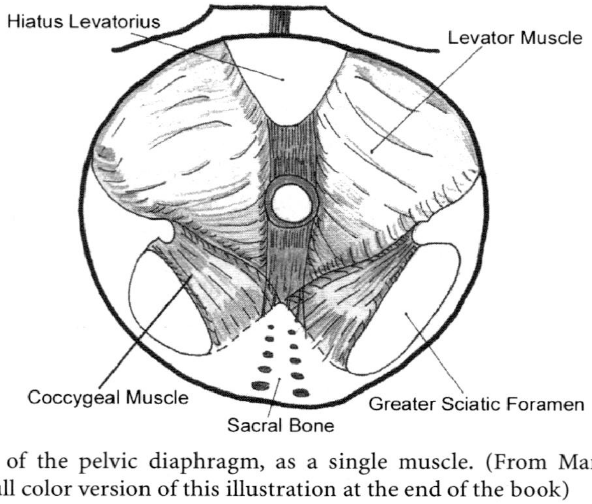

Fig. 2 Muscles of the pelvic diaphragm, as a single muscle. (From Marani 2002, with permission) (full color version of this illustration at the end of the book)

1.2.2.1
Musculus Pubococcygeus

The m. pubococcygeus in humans varies considerably as to division and structure. The most medial fibers originate near the pubic symphysis and are in close apposition to the prostate in the male and to the base of the bladder in the female. Behind the prostate, the medial fibers come in close contact with each other, eventually inserting in the rectum. A difference of opinion exists between anatomists whether these muscular fibers actually cross the midline at this point. Bryce (1923) describes the crossing of fibers within the perineal body, whereas Delbet (1907) and Elftman (1932) describe smooth muscle fibers holding the two opposite pubococcygeus muscles together. The lateral fibers pass backwards, to insert in the rectum, in the anococcygeal ligament and in the coccyx (Elftman 1932).

1.2.2.2
Musculus Puborectalis

There is also a more superficial band of fibers originating near the pubic symphysis, encircling the pelvic viscera as a sling or inserting directly into them. This band is referred to as the puborectalis muscle (Holl 1897; Piersol 1930).

1.2.2.3
Musculus Iliococcygeus

The m. iliococcygeus appears to originate from the tendinous arch of the levator ani. Posteriorly it may originate from the tendinous arch of the fascia pelvis, but anteriorly the "white line" occupies a lower level (it is interesting to see that some authors, for instance Piersol [1930], describe the "white line" as being the tendinous arch of the levator ani). The iliococcygeus is directed downwards, backwards, and inwards, and attaches to the coccyx and the anococcygeal ligament. Its degree of development is variable. It may fail entirely, it may be replaced by fibrous tissue, or it may be strongly developed (Elftman 1932; Thompson 1899; Abitbol 1988). Derry (1907) and Barnes (1921) state that the iliococcygeal muscle overlying the internal obturator muscle is replaced by aponeurotic tissue, when the muscle does not originate from the iliopectineal line but from a lower level. They thus describe the iliococcygeal muscle to originate from the iliopectineal line (i.e., linea arcuata or linea terminalis, being the most medial border of the ilium). However, in most cases it seems to originate from a lower level (the tendinous arches of levator ani or fascia pelvis), and the part covering the internal obturator muscle is thus thin and transparent. If muscular bundles are present, they are frequently separated by membranous intervals, leaving only support as function.

Sometimes another accessory muscle can be found, the m. iliosacralis. This is in fact the posterior part of the iliococcygeus, which can be divided into a ventral and posterior part according to Holl (1897). Usually, only the ventral (iliococcygeus) part persists (Thompson 1899).

1.2.2.4
Musculus Coccygeus

The m. ischiococcygeus or coccygeus muscle is variably described as a part of the levator ani (Quain 1899; Mostwin 1991), or as a single muscle (Waldeyer 1899; Piersol 1930). Evolutionarily, it changed to a fibrous strand when the tail in humans disappeared, forming also most or part of the sacrospinous ligament (Elftman 1932; Thompson 1899; Knauer 1914). Occasionally it may persist as a muscle.

1.3
Embryology

Little is known about the embryology of the human m. levator ani. Power (1948) described the levator ani as the lower part of the rectus abdominis muscle, separated by the in-growth of the pubic bone.

Popowsky (1899) and Gegenbauer (1903) assumed that both the muscles of the urogenital diaphragm and the external anal sphincter are derived from the cloacal sphincter, a circular muscle surrounding the primitive cloaca, and fixed to symphysis, ischial spine, and sacrum. During the development of the perineal fold, the cloacal sphincter is divided into two separate sphincters, the ventral being the sphincter of the urogenital sinus, and the dorsal the external anal sphincter. According to Popowsky (1899), the urogenital diaphragm is derived from the sphincter of the urogenital sinus. A detailed overview of the embryology of the urogenital diaphragm and anal sphincters is beyond the scope of this review: Dorschner et al. (2001) give a detailed overview of the mature anatomy of the urogenital diaphragm, and van der Putte (2005) published an overview of the anal and perineal development.

The development of the pelvic diaphragm is still largely a mystery. One of the authors describing the development of the pelvic diaphragm is Popowsky (1899). Due to extended defects in the material he worked with, he was unable to describe the early stages of the development of the levator ani. In the late stages of development he had to work with, the levator ani was divided in what he called a "pars pubica" and a "pars iliaca," and it could not be separated from the coccygeal muscle. According to Popowsky, an additional argument for lumping these two together was that the coccygeal and levator ani muscles share the same nerve supply (more then a century later, this shared innervation has been demonstrated again by Barber et al. [2002]). This led Popowsky to believe that the levator ani is in fact a differentiation of the coccygeal muscle. Part of the muscle fibers of the coccygeus muscle migrate anteriorly towards the pubic bone, following the lateral borders of the small pelvis. Once these fibers reach the pubic region and attach to the developing bones, they radiate into parallel fibers extending to the pelvic organs. Coming across any muscle derived from the cloacal sphincter, the developing levator ani more or less attaches to these muscles.

Nearly 100 years later, Fritsch and Fröhlich (1994) describe in their article the developmental characteristics of the levator ani in several stages, which will be discussed below.

At the beginning of the fetal period (9th week of development), the *anlage* of the levator ani can already be recognized as consisting of three portions, separated by loose mesenchyme: pubococcygeus, iliococcygeus, and puborectalis. There is a clear border of mesenchyme between the iliococcygeus and coccygeus muscles. The pubococcygeus and puborectalis originate in part from mesenchyme at the pelvic surface of the pubis, lateral to the symphysis. The iliococcygeus is attached to the lower coccygeal vertebrae, which have an acute angle with the sacral vertebrae. In 14-week-old fetuses, the funnel-shaped form of the pelvic diaphragm is already completed (Fritsch and Fröhlich 1994).

In the second trimester of pregnancy (19–28 weeks postconception), the differentiation of connective tissue has proceeded, and the pelvic fascia is clearly outlined. All points of insertion of the levator ani can be recognized at these stages. The pubococcygeus arises from the pubic bone and the anterior part of the obturator fascia. At a level where the tendinous arch of the pelvic fascia fuses with the obturator fascia, the iliococcygeus takes its origin. Differences between male and female pelvic diaphragm now become evident. In the female, a small layer of connective and adipose tissue separates the levator ani from the pelvic organs. This layer is thin and already intermingled with connective tissue. In the male, the levator ani is a well-developed muscle.

During the entire fetal period, no muscle fibers of the levator ani can be seen radiating into the prostate, vagina, or rectum. This is in accordance with studies in adults conducted by Levi et al. (1991), who found the puborectalis, iliococcygeus, and pubococcygeus to have the same origin, never directly connected to the pelvic viscera or anal sphincters. In adults, Dickinson (1889) already could not find any fibers of the levator ani actually inserting in the pelvic viscera.

1.4 Materials and Methods

During prenatal development, four consecutive stages can be discerned (Gribnau and Geijsberts 1981):

1. The presomite stages 1–8
2. The somite stages 9–12
3. The postsomite or organogenetic stages 13–23
4. The fetal stages

From a survey published by Gribnau and Geijsberts (1981), one can conclude that the development of the pelvic floor in humans takes place in the developmental (organogenetic) stages 13–23 (Table 1).

We have chosen the stages 13–23 based on the fact that in stage 14 a definite hind limb bud is present and on Fritsch and Fröhlich's (1994a) description of the presence of the levator ani at 9 weeks postconception (this would be about stage 23). One can expect the pelvic floor to develop in that period. To fully complete the survey, we extended the period of interest to the early fetal period: up to embryos of 50 mm crown-rump length (CRL), 12 weeks postconception.

In the past decennia, nonhuman primates have been used increasingly for research purposes in various scientific fields. Understandably, ethical issues prevent us from unlimited gathering of human embryonic material. Previous authors, especially Gribnau and Geijsberts (1981), have extensively described the similarities in development between different species, making it possible to extrapolate scientific conclusions made in nonhuman primates to humans.

We were lucky to study the development of the pelvic floor in humans (*Homo sapiens*), because an excellent series of sections of human embryos in every different organogenetic stage is held in the collection of the Department of Anatomy and Embryology of the Leiden University Medical Centre. The collection was initiated by the anatomist Dankmeijer and covers a period from the late 1950s to the early 1990s.

Use was made of 38 embryos, cut transversely. Almost all sections were 10 µm thick and fixed in formaldehyde 4%, after which they were stained with hematoxylin-eosin (HE) or Azan.

We were able to select embryos varying in CRL from 9.5 mm CRL (6 weeks postconception) to 50 mm CRL (12 weeks postconception), with small intervals, rendering a detailed overview of the entire organogenetic and early fetal period (Table 2).

Age in CRL is preferred over age in days or developmental stage, because this is the only direct information available in all embryos, allowing for comparison between our data and other studies. Age and stage are extrapolated using the data in Table 1.

Table 1 Embryonic age (O'Rahilly 1979) and crown–rump length (CRL) (Gribnau and Geijsberts1981) of the organogenetic stages in humans

Developmental stage	Age (days)	CRL (mm)
13	28–32	4–6
14	31–35	5–7
15	35–38	7–9
16	37–42	8–11
17	42–44	11–14
18	44–48	13–17
19	48–51	16–18
20	51–53	18–22
21	53–54	22–24
22	54–56	23–28
23	56–60	27–31

Developmental stages are used in the embryonic period

Table 2 Survey of series of sections of human embryos

CRL	Stage	WK	Series	Box	Staining	Sex
9.5 mm	16	3122	56	20	HE	
10 mm	16	6320	163	61	HE	
11 mm	16	5118	113	47	HE	
12 mm	17	1453	45	15	HE	
13 mm	17	4756	98	41	HE	
14 mm	17	5569	144	57	HE	
15 mm	18	5442	136	55	HE	
17 mm	**18**	**4132**	**91**	**35**	**HE**	**Male**
17 mm	**18**	**5032**	**108**	**48**	**HE**	**Male**
17 mm	**18**	**359**	**11**	**2**	**HE**	**Male**
17 mm	**18**	**7435**	**207**	**76**	**HE/Azan**	**Male**
18 mm	19	7904	277	100	HE	
18 mm	19	13	20	3	HE	
18 mm	19	4657	104	34	HE	
20 mm	**20**	**7996**	**297**	**107**	**HE**	
20 mm	**20**	**7262**	**203**	**74**	**HE**	
20 mm	**20**	**7993**	**294**	**104**	**HE**	**Male**
20 mm	**20**	**1423**	**50**	**17**	**HE**	
21 mm	20	5578	146	58	HE	
21 mm	**20**	**2287**	**51**	**17**	**HE**	
21 mm	**20**	**8071**	**305**	**107**	**HE**	
22 mm	21	5121	114	48	HE	
22 mm	21	7489	214	84	HE	
22 mm	**21**	**7832**	**262**	**96**	**HE**	
22 mm	21	7995	296	107	HE	
23 mm	**21**	**548**	**30**	**8**	**HE**	
23 mm	**21**	**5249**	**126**	**51**	**HE**	
23 mm*	**21**	**1256**	**47**	**15**	**HE**	**Male**
25 mm	22	107	7	3	Azan	
25 mm	22	5860	195	71	HE/Azan	
28 mm	23	7629	236	88	HE	
28 mm*	**23**	**1451**	**71**	**24**	**HE**	**Female**
30 mm	**23**	**7652**	**239**	**90**	**HE**	**Female**
34 mm*	**fetal**	**1438**	**61**	**30**	**HE**	**Female**
35 mm	**fetal**	**7608**	**229**	**83**	**HE**	**Female**
42 mm	**fetal**	**2322**	**74**	**28**	**HE**	**Female**
43 mm*	**fetal**	**4409**	**100**	**43/45**	**HE**	**Female**
50 mm*	**fetal**	**3083**	**57**	**21/22**	**HE**	**Female**

In boldface are the embryos described in Sects. 2 and 3 of the text. The other embryos are used as reference, to exclude inter-embryonic variation. Sex typing was only performed when necessary for the description of the embryos. All embryos are cut in 10-μm sections. The embryonic stage is extrapolated from the data presented in Table 1. The reconstructed embryos are marked with an asterisk. *WK* is the abbreviation for work card, with the identification number of the embryo

The embryos and fetuses are described in a declining order of CRL, facilitating an exact identification of the muscles involved in the larger fetuses. Based on characteristics and developmental stage, the muscles can then be identified backwards, in the smaller fetuses and embryos.

1.4.1 Detailed Overview of Series Used

At the Department of Anatomy and Embryology of the Leiden University Medical Centre, large series of embryos of different species have been collected over the years and archived in detail. A small file is kept of important observations and of the preparations made of the embryos for most series. What follows is a description of these files for some of the embryos used. Identification is made by WK-numbers listed in Table 2.

3122 Human embryo with a CRL of 9.5 mm. Originally fixed in alcohol of unknown concentration and after arrival at the department fixed in formaldehyde 4% (1962). Complete with umbilical cord and placenta. Cut in 10-μm sections transversely and stained with HE.

6320 Human embryo with a CRL of 10 mm. First pregnancy of a 25-year-old woman. Fixed according to Bouin (1978), cut in 10-μm sections transversely and stained with HE.

5118 Human embryo with a CRL of 11 mm. Tubal pregnancy. The embryo was immediately fixed after operative removal from the Fallopian tube in formaldehyde 4% (1971). The material appeared in exceptionally good condition. Cut transversely in 10-μm sections and stained with HE.

1453 Human embryo with a CRL of 11 mm. Fixed in alcohol (1957), cut in 10-μm sections and stained with HE.

4756 Human embryo with a CRL of 13 mm. Appearance is slightly macerated. Complete with fetal membranes. Fixed in formaldehyde 4% (1970) after removal of the membranes. Cut transversely in 10-μm sections and stained with HE.

5569 Human embryo with a CRL of 14 mm. Aborted second pregnancy of a 24-year-old woman. Microscopically very slightly macerated. Fixed in formaldehyde 4% (1973) and cut transversely in 10-μm sections. Stained with HE.

5442 Human embryo with a CRL of 15 mm. Macroscopically in excellent condition. Tubal pregnancy. After removal fixed in formaldehyde 4% and cut transversely in 10-μm sections (1973). Stained with HE.

7435 Human embryo with a CRL of 17 mm. Embryo of a 28-year-old female, complete within the embryonic membranes. Fixed in formaldehyde 4% (1987) and cut transversely in 10-μm sections. Stained alternately with HE and Azan.

4657 Human embryo with a CRL of 18 mm. Present in uterus extirpated from a 43-year-old woman. Complete with placenta and fetal membranes. After opening of the membranes, removal of the embryo and fixation in Bouin and alcohol 80% 1 day after surgery (1970). Cut transversely in 10-μm sections and stained with HE.

7996 Human embryo with a CRL of 20 mm. Macroscopically of excellent quality with only minor skin lacerations in the cervical region. Fixed in formaldehyde 4% (1991) and cut transversely in 10-μm sections. Stained with HE.

5578 Human embryo with a CRL of 21 mm. Fixed in formaldehyde 4% (1974) and cut transversely in 10-μm sections. Stained with HE.

7832 Human embryo with a CRL of 22 mm. Fixed in formaldehyde 4% (1990) and cut transversely in 10-μm sections. Stained with HE.

5249 Human embryo with a CRL of 23 mm. Aborted pregnancy of a 29-year-old woman. Fixed in formaldehyde 4%, cut transversely in 10-μm sections and stained with HE.

5860 Human embryo with a CRL of 25 mm. Fixed in formaldehyde 4% (1975) and transversely cut in 10-μm sections. Stained alternately with HE and Azan.

7629 Human embryo with a CRL of 28 mm. Fixed in formaldehyde 4% (1989) and transversely cut in 10-μm sections. Stained with HE.

7652 Human embryo with a CRL of 30 mm. Fixed in formaldehyde 4% (1989) and transversely cut in 10-μm sections. Stained with HE.

1438 Human embryo with a CRL of 34 mm. Fixed in formaldehyde 4% (1957) and transversely cut in 10-μm sections. Stained with HE.

7608 Human embryo with a CRL of 35 mm. Fixed in formaldehyde 4% (1988) and transversely cut in 10-μm sections. Stained with HE.

4409 Female human embryo with a CRL of 43 mm. Fixed in formaldehyde 4% (1969) and cut transversely in 10-μm sections. Stained with HE and Azan alternately.

1.4.2 Sex Typing

In 1961, Lyon outlined the X-inactivation, or what is commonly known as the Lyon hypothesis. It states that:

1. Only one of the X chromosomes is genetically active.
2. The other X chromosome of either maternal or paternal origin undergoes heteropyknosis and is rendered inactive.
3. Inactivation of either the maternal or paternal X chromosome occurs at random among all the cells of the blastocyst on or about the 16th day of embryonic life.
4. Inactivation of the same X chromosome persists in all the cells derived from each precursor cell (Cotran 1999).

The inactive X chromosome can be identified in the nucleus of a cell in interphase. It appears as a darkly staining small mass in contact with the nuclear membrane. It is also known as the Barr body or X chromatin. Barr bodies are present in all somatic cells of normal females, but they are most readily demonstrated in smears of buccal squamous epithelial cells. Barr bodies can also be demonstrated in extraembryonic membranes at about day 12 (presomite stage) of gestation (Miller 1996). X inactivation occurs during the early embryogenesis in a developmentally regulated manner. It occurs first in the extraembryonic trophectoderm and is nonrandom. The paternal X chromosome is exclusively inactivated. Later, in the embryo itself, random inactivation of either the maternal or paternal X chromosome occurs.

Since Barr bodies can thus be identified in embryos from the 16th day of gestation (presomite stage 7), demonstrating these inactive X chromosomes in the embryos we used for our study can reveal the gender of the embryo. For this, we located suitable cells in all embryos (if possible the oral cavity) to look for X chromatin near the nuclear membrane of the cells. Demonstrating a Barr body shows the embryo is female, lacking a Barr body shows the embryo is male. This specific sex typing was only performed in embryos not directly recognizable as male or female by the microscopic appearance of the internal genitalia (i.e., the undoubted presence or absence of a developing uterus, or the appearance of the mesonephric and paramesonephric ducts), only when necessary for the description of the embryos or to study the presence of a sex-type-dependent pelvic diaphragm.

1.4.3 Three-Dimensional Reconstruction

All embryos used in this study were digitally photographed using an Olympus digital 1.2 Megapixel camera (Olympus DP-10, Olympus Optical Company Europe, Hamburg, Germany) mounted on an Olympus light microscope (Olympus BH-2, Olympus Optical Company Europe). The photographs were stored on compact disc at maximum resolution.

The commercially available Surfdriver 3.5.6 software (joint venture of the University of Hawaii, USA and the University of Alberta, Canada; www.surfdriver.com) was used to make three-dimensional reconstructions of the pelvis and pelvic diaphragm of the embryos. With this software, it is possible to trace the outlines of each slice, and then automatically stack the traces for a three-dimensional image. Adjustment of the traced slices is performed manually.

For reconstruction purposes, only the actual muscle volume of the pelvic diaphragm muscles has been reconstructed, without fascia or ligaments. The pelvis has been reconstructed following the contours of the cartilage present, disregarding cartilaginous lines between different developing bones, and disregarding mesenchymal formations. This means that the pelvis as reconstructed in this study appears to be constituted of two ossa coxae, while in fact the os coxa has not yet formed out of the three separate bones (os pubis, ischii, and ilium), which can be identified as such in the actual fetus.

2
Fetal Period

2.1
50 mm (WK 3083), ± 12 weeks

2.1.1
General Features

This fetus with a CRL of 50 mm is about 12 weeks postconception. It is cut transverse in 10-μm sections and stained with HE (see Table 2).

From the appearance of the internal pelvic organs, it is apparent that this is a female fetus, as a uterus and two ovaries can be clearly identified in the sections through the pelvic organs. Figure 3 also shows the other pelvic organs, i.e. the bladder, rectum, and ureters. Together they form a corpus pelvinum that is surrounded by an integument of loose connective tissue and fat, interrupted by seemingly empty spaces caused by the peritoneal folds forming the uterovesical and rectouterine pouch. The peritoneum (visceral and parietal) cannot be identified as such.

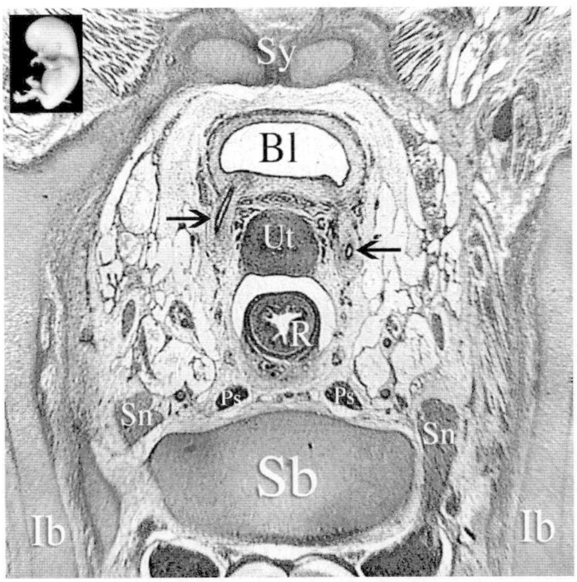

Fig. 3 *Inset*, level of section indicated by a *white line*. *Bl*, bladder; *Ib*, os ilium; *Ps*, plexus sacralis; *R*, rectum; *Sb*, os sacrum; *Sn*, nervus sacralis I; *Sy*, symphysis pubica; *Ut*, uterus; *black arrow*, ureter. The intramural part of the left ureter is clearly visible. Between the uterus and the rectum the rectouterine pouch is identified as a seemingly empty space (cavum Douglasi). On the right side, the first sacral nerve can be seen originating from the spinal cord and passing through the foramen sacralis I. Between the dorsal part of the left os ilium and the os sacrum, the pars lateralis of the os sacrum forms the sacroiliac joint

The development of the bony pelvis is nearly complete. The separately developing bones are outlined in cartilage and no ossification is seen. The left and right os pubis meet, forming the symphysis pubica. Ramus superior and inferior of the os pubis are fully developed forming the foramen obturatum with the corpus and ramus ossis ischii dorsally. In this fetus, the corpus ossis ilii, corpus ossis ischii, and corpus ossis pubis can still be identified individually. Together with the already developed labrum glenoidale they form the acetabulum. The acetabulum can be seen as the center of the os coxae that is formed by the os ilium, os ischii, and the os pubis. The tuber ischiadicum and spina ischiadica of the os ischii are present as mesenchymal condensations and not as cartilaginous tissue. The ala ossis ilii shows a cartilaginous development centrally. The crista iliaca superiorly, the tuberositas iliaca dorsally, and the spina iliaca anterior superior ventrally are not present.

The os sacrum of this fetus is fully developed in cartilage; fusion of the vertebrae is not yet present. Connection of the os sacrum to the os coxae is formed through ligaments attaching the pars lateralis of the os sacrum to the os coxae at the level of the future sulcus paraglenoidalis. These ligg. sacroiliaca interossea can be identified as such. Thus formation of the sacroiliac joint is present. The os coccygis is formed in cartilage and shows fusion of the third to fifth vertebra. It is after reconstruction of the cartilaginous pelvis of this fetus that it became apparent that the typical lordotic angle at the level of the promontorium is not visible and that the os coccygis makes a sharp angle ventrally. Absence of the angle at the level of the promontorium and the sharp angle of the os coccygis seems responsible for the ventral position (see Fig. 13) of the pelvic outlet present in this fetus. This ventral position of the pelvic outlet means that a caudal position in this fetus is in fact a dorsal position in the adult situation.

2.1.2 Musculus Obturatorius Internus

The m. obturatorius internus forms an important landmark in the identification of the muscles of the pelvic diaphragm. In adults, it is a large and clearly outlined muscle covering the os ischii and foramen obturatum. It originates from the ramus superior ossis pubis up to the sacroiliac joint, the ramus inferior ossis pubis, and the ramus ossis ischii (surrounding the margins of the foramen obturatum). It runs dorsally through the incisura ischiadica minor, leaving the pelvic cavity to insert at the medial surface of the trochanter major femoris proximal to the fossa trochanterica.

In this fetus, the m. obturatorius internus is in all respects comparable to the adult situation. It can be seen originating from the medial aspect of the ramus superior ossis pubis, the ramus inferior ossis pubis and ramus ossis ischii covering the foramen obturatum and corpus ossis ischii. It can also be seen leaving the pelvic cavity to insert in the trochanter major femoris. Laterally the fetal m. obturatorius internus is separated from the foramen obturatum by a thin layer of connective

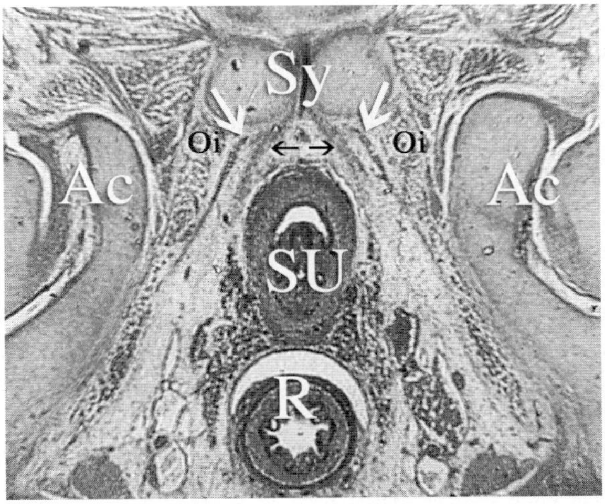

Fig. 4 *Ac*, acetabulum; *Oi*, m. obturatorius internus; *SU*, sinus urogenitalis; *Sy*, symphysis pubica; *R*, rectum; *black arrow*, endopelvic fascia and pubovesical ligament; *white arrow*, arcus tendineus levatoris ani (ventral part near os pubis). Note the close relationship of the ATLA and the endopelvic fascia together with the pubovesical ligament, anchoring the urogenital sinus to the os pubis and pelvic diaphragm

tissue; the membrana obturatoria. At the other side of the membrana obturatoria the m. obturatorius externus can be identified clearly.

The nervus obturatorius is leaving the pelvis, nearly completely surrounded by the m. obturatorius internus, through the canalis obturatorius. Only the superior margin of the canalis obturatorius is formed by cartilage, i.e., the ramus superior ossis pubis.

Just below the canalis obturatorius there is a dense band of connective tissue almost immediately covering more than half of the medial aspect of the m. obturatorius internus and extending farther dorsally from the inner opening of the canalis obturatorius. This is the developing arcus tendineus levatoris ani (ATLA). Ventrally it is fixed to the inner aspect of the ramus inferior ossis pubis just laterally of the symphysis pubica. Here it comes in to close contact with a small, dense band of connective tissue arising from the middle of the symphysis extending around the urogenital sinus; the developing pubovesical ligament and endopelvic fascia (see Fig. 4). Dorsally the ATLA disappears in loose connective tissue. Following this arcus caudally it smoothly changes in muscular fibers: the m. levator ani.

2.1.3 Musculus Levator Ani

At both the left and right side of the pelvis, the levator ani develops and shows a gradually increasing thickening as it progresses caudally. Its muscular fibers show an almost right-angled orientation to the ATLA. From the moment the levator ani originates, it is clear that it is a single muscle. Near its origin at the level of the ATLA, it comes into close contact with the inner aspect of the ramus inferior ossis pubis, only separated from it by a small border of connective tissue. Dorsally the levator ani extends over about half of the medial aspect of the m. obturatorius internus, fading away in connective tissue. As soon as the levator ani originates from the ATLA it is separated from the m. obturatorius internus by a narrow space. On the medial side of the levator ani, a small border of connective tissue (no fat) separates the muscle from the urogenital organs.

Progressing caudally, the thickness of the levator ani increases, especially on the ventral side. The distance between the ramus inferior ossis pubis and the ventral side increases and the developing arcus tendineus levator ani starts to extend dorsally to the most dorsal part of the corpus ossis ischii, covering the m. obturatorius completely.

At the lower level of the foramen obturatum, the levator ani has developed to a muscle completely covering the m. obturatorius internus, almost fixed to the dorsal end of the corpus ossis ischii (the developing spina ischiadica). The decreasing ventral–dorsal length of the m. obturatorius internus also contributes to this covering. On the ventral side the levator ani comes into close contact with the hind side of the urogenital sinus, but is not fixed to it. On the dorsal side, the levator ani extends farther away from the rectum. At this level, a third muscle appears at the most dorsal part of the corpus ossis ischii. This is the m. coccygeus that will be discussed later on (see Fig. 5).

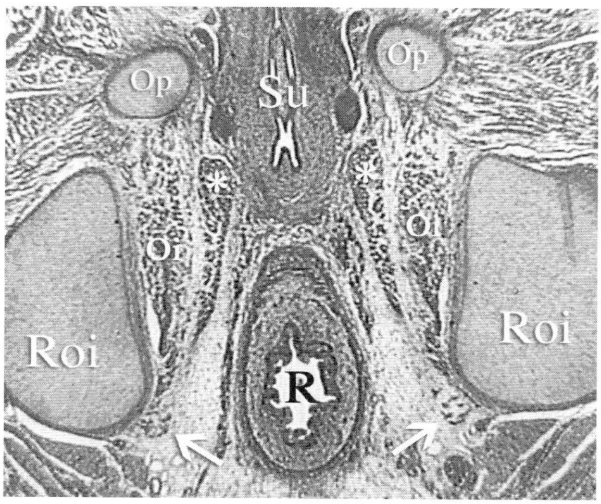

Fig. 5 *Oi*, m. obturatorius internus; *Op*, os pubis (ramus inferior); *R*, rectum; *Roi*, ramus ossis ischii; *Su*, sinus urogenitalis; *white arrow*, m. coccygeus; *, m. levator ani. Note that the m. levator ani lays completely separated from the sinus urogenitalis, the m. obturatorius internus, and rectum. Dorsally, it has a continuation towards the ramus ossis ischii. This continuation is in fact the ATLA. It inserts on the surface of the os ischii together with the m. obturatorius internus and the origin of the m. coccygeus

Progressing even further caudally through the serial sections and coming behind the urogenital sinus, it appears that the levator ani has retracted from its attachment to the corpus ossis ischii. The dorsal end of the levator ani now lies more closely to the rectum without touching it and away from the corpus ossis ischii. The ventral side of the levator ani makes contact with a dens structure of connective tissue that crosses the midline just in front of the rectum to contact the ventral side of the levator ani on the other side of the pelvis. This structure is identified as the developing centrum tendineum (see Fig. 6). At the level of the developing centrum tendineum no muscle fibers can actually be seen crossing the midline.

Further on, behind the centrum tendineum, the sphincter complex of the rectum becomes visible with a m. sphincter ani externus and internus. The ventral side of the levator ani, which now lies adjacent to the rectum, is fixed to the m. sphincter ani externus, i.e., the deep part of the external sphincter. The rest of the levator ani is still one single muscle on the left and right side of the rectum without being attached to it (Fig. 7).

Just behind the rectum, the left and right levator ani near each other in the midline. They are still separated by a structure of dense connective tissue that connects the left and right levator ani to the sphincter complex of the rectum (Fig. 8). This structure is the developing ligamentum or raphe anococcygeum. It does not contain muscle fibers. Farther caudally, the left and right levator ani disappear still as separate muscles into the ligamentum anococcygeum that can be identified until its insertion in the tip of the os coccygis.

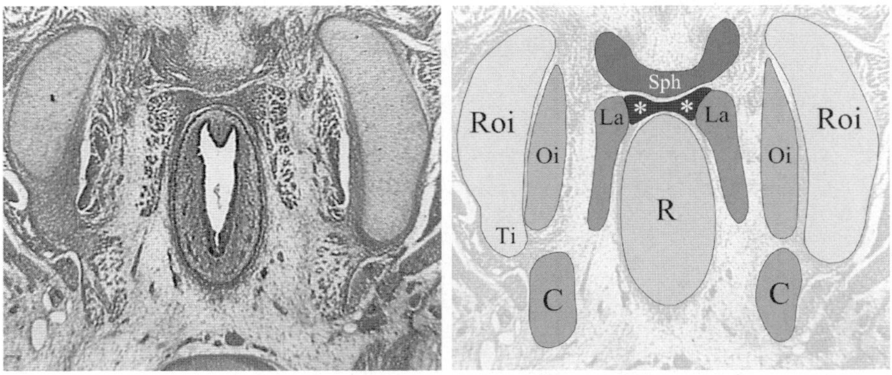

Fig. 6 Section at a level between the sinus urogenitalis and rectum. *C*, m. coccygeus; *La*, m. levator ani; *Oi*, m. obturatorius internus; *R*, rectum; *Roi*, ramus ossis ischii; *Sph*, m. sphincter ani externus (superficial par); *Ti*, tuber ischiadicum; *, m. sphincter ani externus (deep part). The levator ani is attached to the deep part of the external sphincter at the level of the perineum. The levator ani is nowhere attached to the rectum. This picture shows that the external sphincter is in fact the only muscle present in the midline of the pelvic diaphragm. Left and right levator ani are attached to each other by the deep part of the external sphincter, neither of them actually crosses the midline nor do they show any direct contact

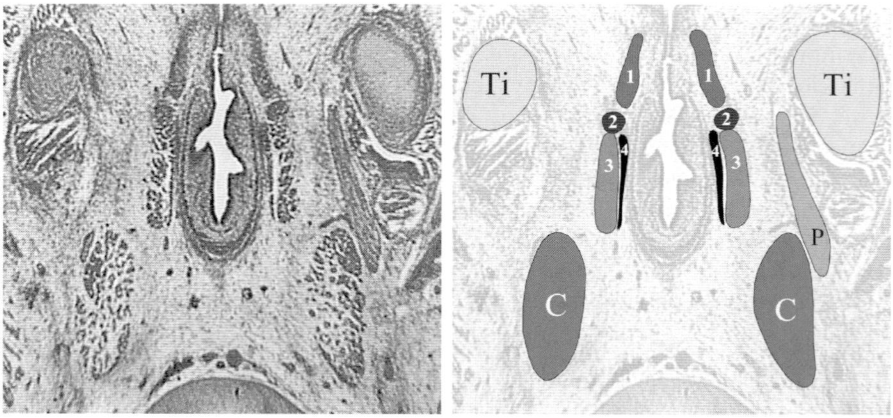

Fig. 7 Section at a level right through the anal canal. *C*, os coccygeus; *P*, nervus pudendus; *Ti*, tuber ischiadicum; *1*, superficial part of m. sphincter ani externus; *2*, deep part of m. sphincter ani externus; *3*, m. levatoris ani; *4*, developing longitudinal ligament. Again it becomes clear that the levator ani is attached to the deep part of the external sphincter as it passes alongside the rectum. There is no attachment to the rectum nor the internal sphincter. The left m. coccygeus shows its subdivision in a ventral and dorsal part. The dorsal part becomes the ligamentum sacrospinale (see Figs. 8 and 9)

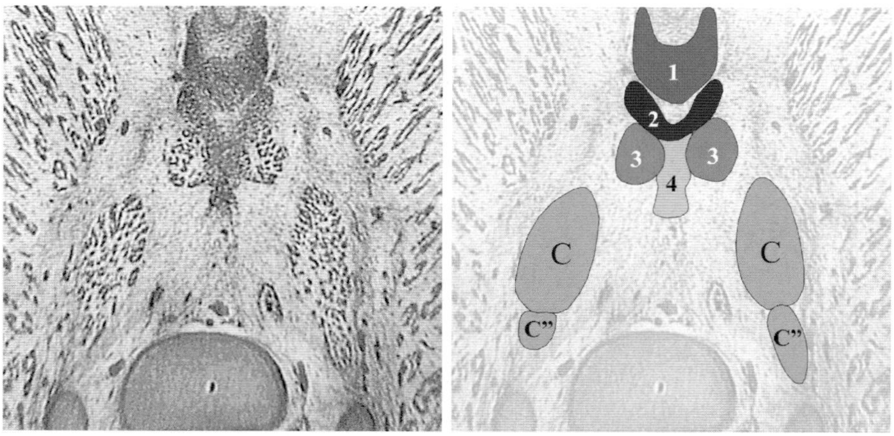

Fig. 8 Section just behind the rectum and in front of the os coccygis. *C*, m. coccygeus; *C"*, dorsal part of m. coccygeus (future ligamentum sacrospinale); *1*, superficial part of m. sphincter ani externus; *2*, deep part of m. sphincter ani externus; *3*, m. levator ani; *4*, ligamentum anococcygeum. The left and right levator ani are attached to the deep part of the external sphincter and the ligamentum anococcygeum. They are not directly attached to each other. The m. coccygeus is divided in a ventral and dorsal part

2.1.4 Musculus Coccygeus

At the surface of the most dorsal part of the corpus ossis ischii, above the developing tuber ischiadicum, the origin of the m. coccygeus can be identified. This is also the point where the ATLA is attached to the corpus ossis ischii and just dorsal of the insertion of the m. obturatorius internus onto the corpus ossis ischii. This point is the future spina ischiadica. The origin of the m. coccygeus extends to the developing tuber ischiadicum.

More caudally, the m. coccygeus starts to extend dorsally and expand towards the os sacrum. At a level just below the developing tuber ischiadicum, it becomes apparent that the m. coccygeus is divided into a dorsal and ventral part by a structure that appears to be a nerve and blood vessels (Fig. 9). When following this structure, it appears to be a branch from the nervus pudendus that runs medial from the developing tuber ischiadicum and lateral from the m. coccygeus. This branch has to be (one of) the innervating nerve of the m. coccygeus. Nearly all the

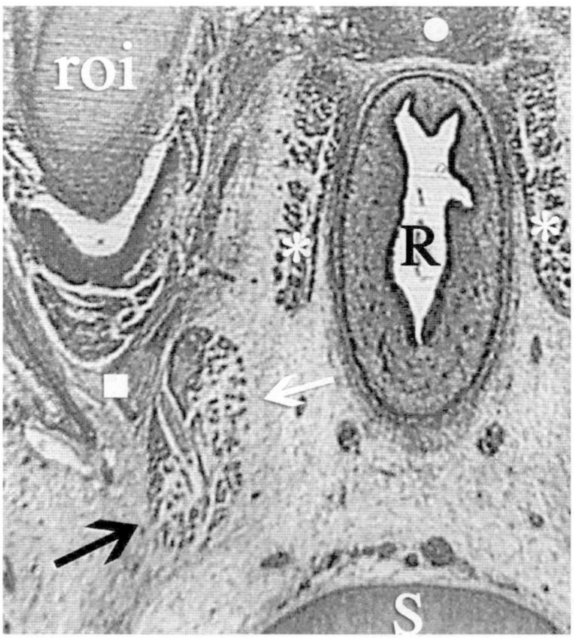

Fig. 9 Close-up of Fig. 6. *R*, rectum; *roi*, ramus ossis ischii; *S*, os sacrum; *white dot*, external anal sphincter; *, m. levator ani; *white square*, nervus pudendus; *black arrow*, ligamentum sacrospinale (dorsal part m. coccygeus); *white arrow*, m. coccygeus (ventral part). The m. coccygeus (here on the *left side*) is divided into a ventral and a dorsal part by a structure that appears to be a branch of the nervus pudendus. Following this division, it becomes apparent that the dorsal part is the future ligamentum sacrospinale. This would explain the close relationship between the ligamentum sacrospinale and the m. coccygeus in adult humans

way up to the insertion of the m. coccygeus onto the os sacrum, this division of the muscle in a ventral and dorsal part can be seen (Fig. 8). There can be no other explanation than that this dorsal part of the m. coccygeus is in fact the future ligamentum sacrospinale. This would explain the close relationship of the m. coccygeus with the ligamentum sacrospinale in adults. Following the insertion of the m. coccygeus onto the os sacrum more caudally, the dorsal part disappears after its insertion onto the os sacrum. The ventral part continues to travel caudally towards what appears to be the last sacral vertebra and the first coccygeal vertebra. The insertion of the ventral part of the m. coccygeus into the os coccygis cannot be seen. The muscle is lost in dense connective tissue (lig. intervertebrale) before ending onto the os coccygis.

The ventral part of the m. coccygeus travels towards the midline, following it caudally through the fetal pelvis, where it points towards the future ligamentum anococcygeum. The m. coccygeus can also be traced much more caudally then the m. levator ani and it never comes into contact with the levator ani. There is always a distinct border of loose connective tissue between the dorsal end of the levator ani and the ventral end of the m. coccygeus.

2.1.5 Musculus Sacrococcygeus Ventralis

In front of the os sacrum on the medial aspect of the m. coccygeus, another muscle is identified in this fetus, both on the left and the right side (Fig. 10). It originates medial on the inner aspect of the os sacrum and can be followed caudally towards the os coccygis, where it inserts on the lateral aspect of the last coccygeal vertebrae. This muscle is the m. sacrococcygeus. Its muscular fibers are directed in such a way that the function of this muscle is to move the os coccygis ventrally: "the tail between the legs." As the os coccygis in humans can be seen as a rudimentary tail, it is likely that the m. sacrococcygeus has disappeared in most cases. Only in some anatomic handbooks (Sobotta, Atlas of Human Anatomy, 1982) is the m. sacrococcygeus described as an existing muscle, leaving it undescribed in others (Gardner et al. 1986). The m. sacrococcygeus is a rudimentary muscle, variably present in adult humans.

In between the left and right m. sacrococcygeus, just in front of the arteria and vena sacralis mediana, a structure of connective tissue appears that seems to originate from the midline of the os sacrum. It inserts on the os coccygis and ligamentum anococcygeum. This has to be either the ligamentum sacrococcygeum ventrale or continuing fibers of the ligamentum anococcygeum (Fig. 10).

2.1.6 Three-Dimensional Reconstruction Oblique View from Above on the Pelvic Diaphragm

Figure 11 illustrates the full reconstruction of the fetal pelvis shown from above. Figure 12, illustrating the reconstruction is taken from above, shows the (right) pelvic diaphragm from the inside of the pelvis. In the lower middle is the os sacrum. In the upper left corner the symphysis pubica.

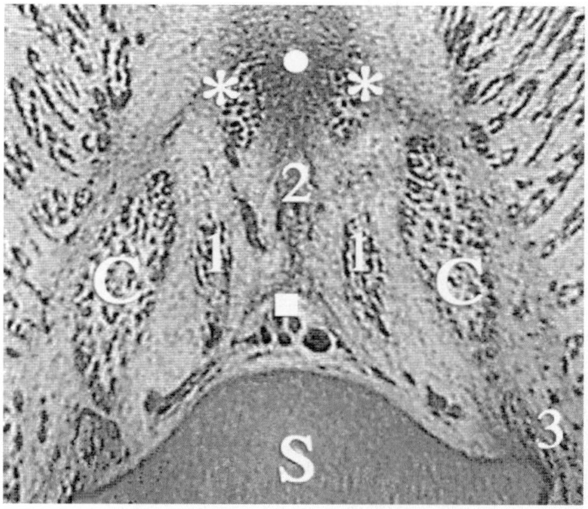

Fig. 10 Section just in front of the tip of the os coccygis. *C*, m. coccygeus; *S*, os sacrum; *white dot*, ligamentum anococcygeum; *white square*, a/v sacralis mediana; *, m. levator ani; *1*, m. sacrococcygeus ventralis; *2*, ligamentum sacrococcygeum ventrale; *3*, future ligamentum sacrospinale. The ventral part of the m. coccygeus runs farther towards the os coccygis. The dorsal part (ligamentum sacrospinale) has just inserted onto the os sacrum. The left and right levator ani are attached to the ligamentum anococcygeum but not to each other

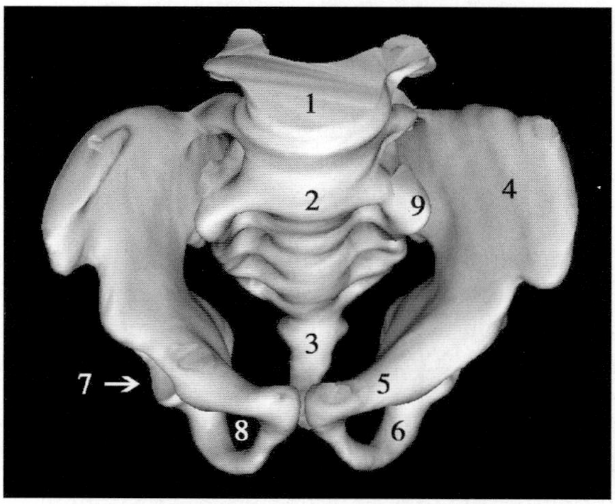

Fig. 11 Overview of the fetal bony pelvis. *1*, last lumbar vertebra; *2*, first sacral vertebra; *3*, os coccygis; *4*, ala ossis ilii; *5*, ramus superior ossis pubis; *6*, ramus inferior ossis pubis; *7*, acetabulum; *8*, foramen obturatum; *9*, pars lateralis of os sacrum. On the *left and right side* of the first sacral vertebra (*2*) the pars lateralis (*9*) is visible, together with *4* forming the sacroiliac joint

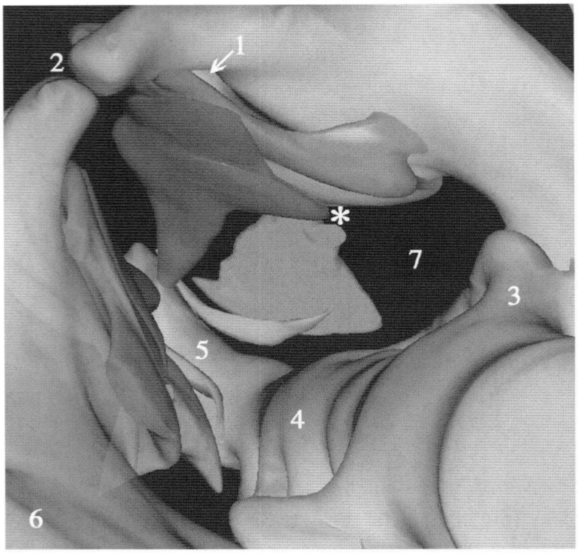

Fig. 12 View from above on the pelvic diaphragm. *1*, canalis obturatorius; *2*, symphysis pubica; *3*, pars lateralis ; *4*, os sacrum; *5*, os coccygis; *6*, os ilium; *7*, foramen ischiadicum majus; *, future spina ischiadica; *red*, m. obturatorius internus; *blue*, m. levator ani; *green*, m. coccygeus; *light blue*, m. sacrococcygeus ventralis. For illustrative purposes, the pelvic viscera (sinus urogenitalis and rectum) have been left out. The m. levator ani closes the ventral half of the pelvic outlet. Ventral to dorsal, it completely covers the m. obturatorius internus. The m. coccygeus closes the dorsal half of the pelvic outlet and has the same size as the m. levator ani (full color version of this illustration at the end of the book)

The left and right corpus ossis pubis meet in the midline, forming the symphysis pubica (*2*). The rectum and sinus urogenitalis, together forming the corpus pelvinum, have been left out.

The reconstruction of the m. obturatorius internus is shown in red. The ventral to dorsal extension from corpus ossis pubis to the corpus ossis ischii, where it leaves the pelvis to insert in the trochanter major of the femur, becomes visible. Between the cranial edge of the m. obturatorius internus and the pelvis the canalis obturatorius appears (*1*).

The reconstruction of the m. levator ani is shown in dark blue. It covers the m. obturatorius internus almost completely. The oblique border between the levator ani and the obturatorius internus is in fact the ATLA, running from the inner aspect of the ramus inferior ossis pubis to the later originating spina ischiadica. The ATLA is the very broad origin of the levator ani in contrast to the narrow insertion on the os coccygis. This reconstruction of the m. levator ani also makes it distinctly clear that the levator ani consists of a single muscle that covers the frontal half of the pelvic outlet and has a somewhat triangular shape.

The m. coccygeus is reconstructed in green. Its origin is at the level of the spina ischiadica, which is formed later, and its insertion is on the surface of the last sacral and first coccygeal vertebra. The m. coccygeus has somewhat the same shape and volume as the m. levator ani and covers the dorsal half of the pelvic outlet.

The m. levator ani and the m. coccygeus are two separate muscles, distinguishable from each other by a small space between them, together closing the pelvic outlet.

The m. sacrococcygeus ventralis is reconstructed as an example of a rudimentary muscle present in this fetus, shown in light blue. It does not cover any part of the actual pelvic outlet and is therefore not considered to belong to the pelvic diaphragm.

2.1.7 Three-Dimensional Reconstruction Frontal and Caudal Views of the Pelvic Diaphragm

Figure 13 shows a complete frontal picture of the reconstruction. It immediately becomes clear that the pelvic outlet of this 50-mm fetus is projected in the frontal plain, thus pointing forwards. This is in striking contrast to the more backward-oriented pelvic outlet in adult humans. Formation of the angle at the level of the promontorium and the disappearing of the ventral angle between os sacrum and os coccygis is responsible for this backward shifting of the pelvic outlet towards

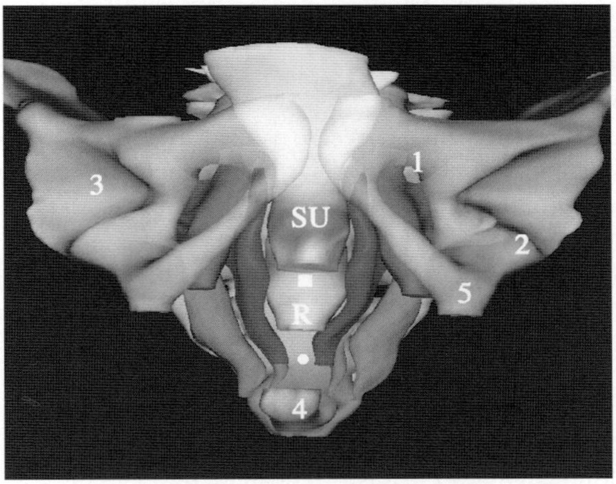

Fig. 13 Frontal view of the pelvic diaphragm. *1*, canalis obturatorius; *2*, m. obturatorius internus leaving pelvis to insert on femur; *3*, acetabulum; *4*, os coccygis; *5*, ramus ossis ischii; *SU*, sinus urogenitalis; *R*, rectum; *red*, m. obturatorius internus; *blue*, m. levator ani; *green*, m. coccygeus; *light blue*, m. sacrococcygeus ventralis. The m. obturatorius internus completely covers the foramen obturatum except for the canalis obturatorius. Note the lateral position of the m. coccygeus in regard to the m. levator ani. The left and right m. levator ani show curving around the sinus urogenitalis and rectum nearing each other at the level of the future centrum tendineum (*white square*) and ligamentum anococcygeum (*white dot*) (full color version of this illustration at the end of the book)

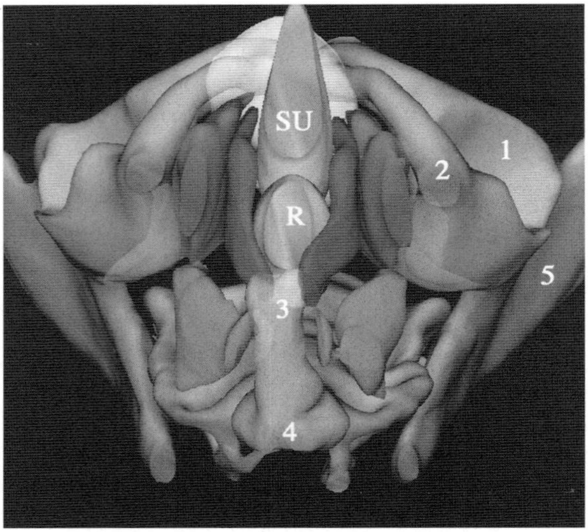

Fig. 14 View of the pelvic diaphragm from below. *1*, acetabulum; *2*, ramus ossis ischii; *3*, os coccygis; *4*, last sacral vertebra; *5*, os ilium; *SU*, sinus urogenitalis; *R*, rectum; *red*, m. obturatorius internus; *blue*, m. levator ani; *green*, m. coccygeus; *light blue*, m. sacrococcygeus ventralis. Note the four-quadrant formation of the pelvic diaphragm (full color version of this illustration at the end of the book)

maturity. The sharp angle of the os coccygis to the os sacrum in this fetus is clearly visible in Fig. 14.

The m. obturatorius internus is reconstructed in red. It covers the foramen obturatum completely except for a small part, which forms the canalis obturatorius (Fig. 13). The m. obturatorius internus in this 50-mm fetus in fact resembles the adult situation.

Again the m. levator ani is shown in dark blue. It runs from the os pubis (corpus ossis pubis, lower margin of the ramus inferior ossis pubis) to the os coccygis. The left and right m. levator ani surround the corpus pelvinum, which is reconstructed in yellow.

The left and right levator ani are each separate muscles and are curved inward and towards each other at a level between the sinus urogenitalis and the rectum and between the rectum and the os coccygis. The point between the sinus urogenitalis and the rectum will form the centrum tendineum as the developing left and right levator ani will eventually meet in the midline. Behind the rectum, in front of the os coccygis, the left and right levator ani will eventually end at the level of the developing ligamentum anococcygeum. Thus closure of the pelvic outlet in the midline by the two halves of the pelvic diaphragm will eventually take place at the level of the centrum tendineum and the ligamentum anococcygeum. However, closure of the pelvic diaphragm in the midline is not yet present in this fetal stage. The insertion of the m. levator ani onto the surface of the os coccygis is not yet completed. A small gap is still present, formed by connective tissue that is not reconstructed.

The m. coccygeus, reconstructed in green, appears to be much thicker than the m. levator ani. In this early fetal period, the m. coccygeus appears to have a much more dominant role in the formation of the pelvic diaphragm as it has in later adult life. The position of the m. coccygeus is more lateral in relation to the m. levator ani. The two muscles show what looks like a roof tile formation.

A close look at Fig. 14 suggests that the pelvic diaphragm consists of four quadrants. The ventral left and ventral right quadrant are filled by the left and right m. levator ani, respectively and the dorsal quadrants by the m. coccygeus (see also Marani 2002).

2.2 43 mm (WK 4409) and 42 mm (WK 3222), ± 10 weeks

2.2.1 General Features

These fetuses with a CRL of 43 mm and 42 mm, respectively, are approximately 10 weeks postconception. They are cut transverse in 10-µm sections and stained with HE (see Table 2).

From the appearance of the internal pelvic organs, it is apparent that these fetuses are female, as a uterus (fused paramesonephric ducts) and two ovaries can be clearly identified in the sections through the pelvic organs. Figure 15 gives an overview of one of the sections through the pelvic organs in the 43-mm fetus. The level of the section taken is just above the ramus superior ossis pubis. On the left side, the m. psoas major and the m. iliacus can be seen leaving the pelvic cavity. The urachus (future bladder) can be identified, as well as the uterus and the colon sigmoideum. Beside the colon sigmoideum the ureters are visible. The corpus pelvinum is embedded in connective tissue with mesenchyme and blood vessels. There seems to be less fat present than in the 50-mm fetus. The uterovesical and rectouterine pouch are visible. On both sides, the external iliac vessels are prominently visible beside the corpus pelvinum.

The development of the bony pelvis is nearly complete. The separately developing bones are all outlined in cartilage and no ossification can be seen. At this stage, the corpus ossis ilii, the corpus ossis ischii and the corpus ossis pubis can be identified individually. Yet in contrast to the 50-mm fetus, several dense centers of developing cartilage can be seen and the volume of the completed cartilage seems to be less than in the 50-mm fetus. This is especially striking at the level between the os ilium and the os ischii and at the level of the developing labrum glenoidale. A thick layer of tissue is surrounding the developing ala ossis ilii. This is developing cartilage (Figs. 15 and 16). The tuber ischiadicum and spina ischiadica of the os ischii are present as mesenchymal condensations and not as cartilaginous tissue. The crista iliaca superiorly, the tuberositas iliaca dorsally and the spina iliaca anterior superior ventrally have not yet developed.

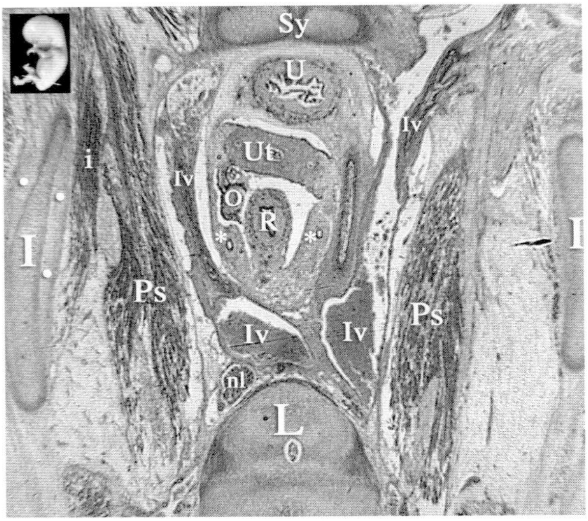

Fig. 15 *Inset*: level of section indicated by *white line*. *I*, os ilium; *i*, m. iliacus; *Iv*, iliac vessel; *L*, fifth lumbar vertebra; *nl*, lymph node; *O*, ovary; *Ps*, m. psoas major; *R*, sigmoid; *Sy*, symphysis pubica; *U*, urachus (bladder); *Ut*, uterus (fused paramesonephric ducts centrally and two mesonephric ducts laterally; *, ureter; *white dot*, developing cartilage. The left os ilium is surrounded by a thick layer of developing cartilage

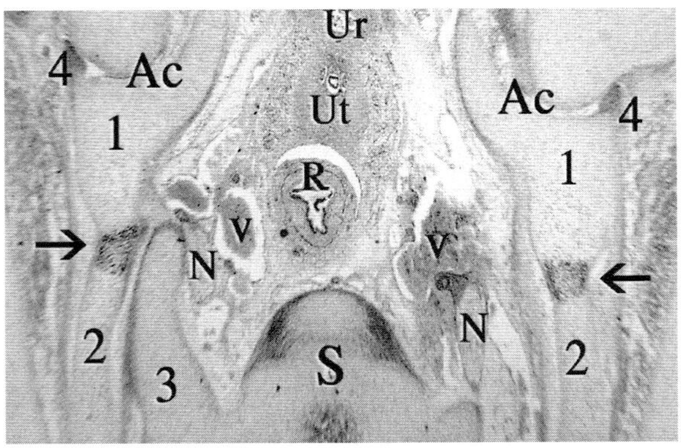

Fig. 16 *Ac*, acetabulum; *N*, first sacral nerve; *Ur*, urachus; *Ut*; uterus; *R*, rectum; *S*, os sacrum (first vertebra); *V*, iliac vein; *black arrow*, border between os ilium and ischii; *1*, os ischii; *2*, os ilium; *3* pars lateralis os sacrum; *4*, labrum glenoidale. Around the os ilium, at the level of the *black arrow*, the thick layer of developing cartilage is clearly identified

The os sacrum of the fetuses is fully developed in cartilage; fusion of the vertebrae is not yet present. Connection of the os sacrum to the os coxae is formed through ligaments attaching the pars lateralis of the os sacrum to the os coxae. These ligg. sacroiliaca interossea can be identified as such. The os coccygis is formed in cartilage and shows fusion of the third to fifth vertebra.

After reconstruction of the cartilaginous pelvis of the 43-mm fetus, it becomes apparent, like the 50-mm fetus, that the typical angle at the level of the promontorium is not visible and that the os coccygis makes a sharp angle ventrally. Absence of the angle at the level of the promontorium and the sharp angle of the os coccygis causes the ventral position of the pelvic outlet present in this fetus.

2.2.2 Musculus Obturatorius Internus

At this stage, the m. obturatorius internus itself does not completely cover the foramen obturatum. Especially at the level of the os pubis the developing muscle fibers do not reach the surface of the inner aspect of the corpus ossis pubis, i.e., the ramus superior and inferior ossis pubis. The fascia of the m. obturatorius internus, however, is attached to the corpus ossis pubis (Fig. 17). Dorsally the m. obturatorius internus reaches as far as it does in the 50-mm fetus. It reaches the most dorsal part of the corpus ossis ischii and can be seen leaving the pelvis to insert onto the trochanter major femoris. Compared to the 50-mm fetus, the m. obturatorius internus is less developed on its ventral and cranial side.

At its lateral side the m. obturatorius is covered by the membrana obturatoria. The membrana obturatoria, however, is thicker than in the 50-mm fetus. It appears to be less condensed at the age of 10 weeks in comparison to the age of 12 weeks.

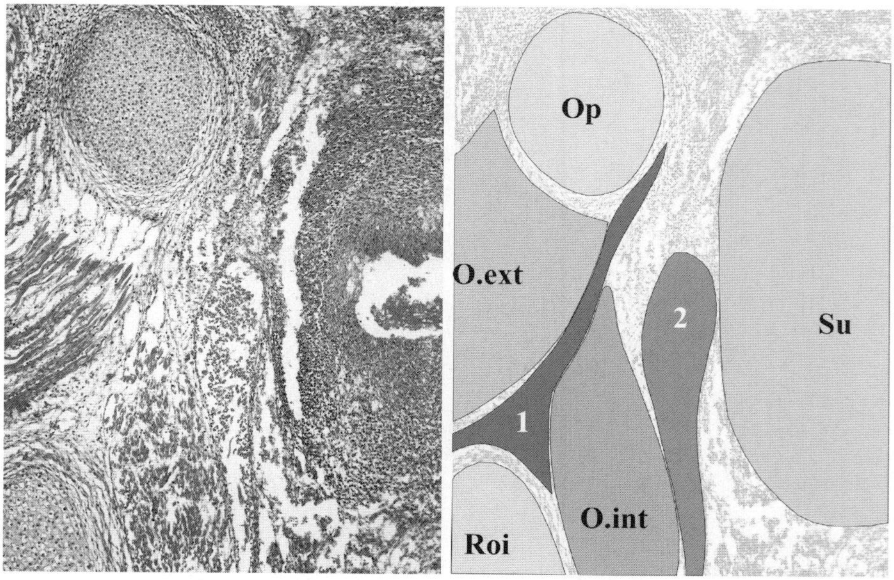

Fig. 17 *O.ext*, m. obturatorius externus; *O.int*; m. obturatorius internus; *Op*, ramus inferior ossis pubis; *Roi*, ramus ossis ischii; *Su*, sinus urogenitalis; *1*, membrana obturatoria; *2*, m. levator ani. The m. obturatorius internus does not reach the os pubis; attachment is made by fascia. The membrana obturatoria is thicker than in the 50-mm fetus

On the other side of the membrana obturatoria the m. obturatorius externus can be identified (Fig. 17).

The nervus obturatorius can be seen leaving the pelvis just below the ramus superior ossis pubis. The nerve, however, is not completely surrounded by the m. obturatorius internus. The muscle starts to appear just below the level of the nervus obturatorius, not present on the ventral and dorsal side of the nervus obturatorius. The canalis obturatorius as such has thus not completely developed.

Just below the level of the symphysis, on the inner aspect of the ramus inferior ossis pubis, a dense band of connective tissue can be seen. It runs towards the urachus and eventually radiating around it. This is the developing pubovesical ligament covered by the endopelvic fascia.

2.2.3 Musculus Levator Ani

On the medial aspect of the m. obturatorius internus, just below the nervus obturatorius, the m. levator ani can be identified. It originates at the inner aspect of the ramus inferior ossis pubis separated by a small border of connective tissue. The muscle's origin is between the m. obturatorius internus and ligamentum pubovesicalis (ventral part of the arcus tendineus fascia pelvis) (Fig. 18). A clear arcus tendineus levatoris ani (ATLA), as in the 50-mm fetus, cannot be identified.

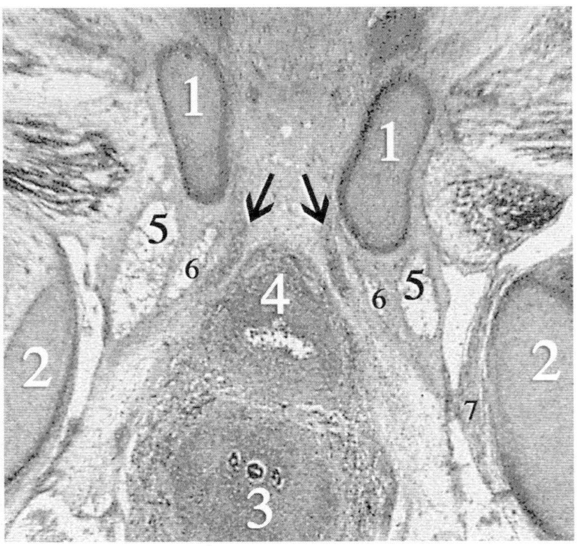

Fig. 18 *1*, ramus inferior ossis pubis; *2*, ramus ossis ischii; *3*, uterus; *4*, bladder; *5*, m. obturatorius internus; *6*, m. levator ani; *7*, nervus obturatorius; *black arrow*, ligamentum pubovesicalis. A dense band of connective tissue runs from the inner aspect of the os pubis alongside the bladder, partially overlying the m. levator ani. This is the arcus tendineus fascia pelvis, i.e., the ligamentum pubovesicalis

Instead a somewhat broader, less condensed strand of connective tissue is present near the origin of the m. levator ani.

As the m. levator ani originates, it covers approximately half of the medial aspect of the m. obturatorius internus and is separated from it by a small band of connective tissue, formed by the individual fascia of the two muscles. As it originates, it is immediately clear that the m. levatoris ani is one muscle. In contrast to the 50-mm fetus, the m. levator ani keeps a close relationship with the m. obturatorius internus as it progresses towards the os coccygis. Behind the sinus urogenitalis, it starts to separate, but keeps close contact dorsally until the m. obturatorius internus disappears as it inserts onto the ramus ossis ischii. The medial aspect of the m. levator ani is separated from the corpus pelvinum by loose connective tissue.

Progressing caudally, the thickness of the m. levator ani starts to increase, especially on the ventral side. The ventral to dorsal length increases and at the lower level of the foramen obturatum the m. levator ani has developed to a muscle almost completely covering the m. obturatorius internus. The decreasing ventral–dorsal length of the m. obturatorius internus also contributes to this partial covering. The covering of the m. obturatorius internus at this stage is not as complete as it is in the 50-mm fetus. Especially on the dorsal side, near the corpus ossis ischii, the m. levator ani does not completely cover the m. obturatorius internus. Complete covering starts to appear near the level of the ramus ossis ischii as the m. obturatorius internus starts to decrease as it inserts on the surface of the ramus ossis ischii. At this level, a third muscle starts to appear. This is the m. coccygeus that will be discussed later on. The levator ani is not directly attached to the dorsal end of the corpus ossis ischii. Attachment is made by connective tissue. This is the ATLA, which is less developed near the ramus ossis ischii and the future spina ischiadica than it is in the 50-mm fetus.

On the ventral side, the distance between the ramus inferior ossis pubis and the m. levator ani increases as it passes alongside the sinus urogenitalis. The m. levator ani lies immediately adjacent to the sinus urogenitalis, but is not attached to it. On the most ventral side, the m. levator ani is attached to a thick layer of connective tissue extending towards the medial aspect of the sinus urogenitalis, the inner aspect of the ramus inferior ossis pubis and the ventral side of the m. obturatorius internus. This is at the level of the urogenital diaphragm (Fig. 19). Just behind the sinus urogenitalis, the deep part of the m. sphincter ani externus appears in front of the m. levator ani. The m. levator ani has a very close relationship with this part of the m. sphincter ani externus, but is not attached to it. In between the sinus urogenitalis and the rectum, a clear connection between the left and right m. levator ani is not present. This is in contrast to the 50-mm fetus, where a developing centrum tendineum can be identified.

Passing around the rectum, the m. levator ani keeps a close relationship with the deeper part of the m. sphincter ani externus, but remains unattached to it. The m. levator ani is always separated from the rectum by a small layer of connective tissue. Behind the rectum the left and right m. levator ani start to come towards

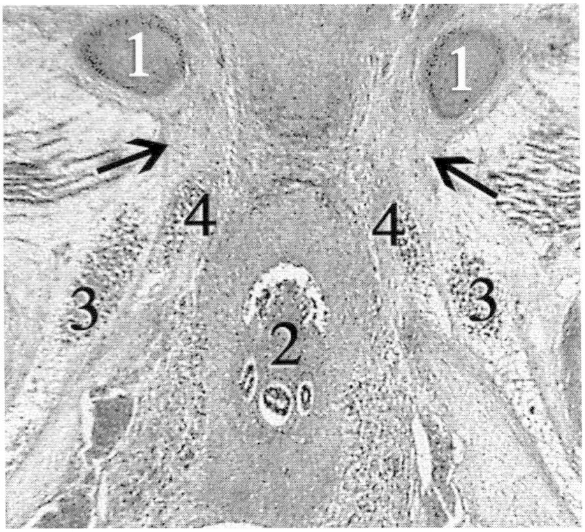

Fig. 19 *1*, ramus inferior ossis pubis; *2*, sinus urogenitalis; *3*, m. obturatorius internus; *4*, m. levator ani. *Black arrow*, the m. obturatorius internus and levatoris ani are separated from the ramus inferior ossis pubis by a thick layer of connective tissue

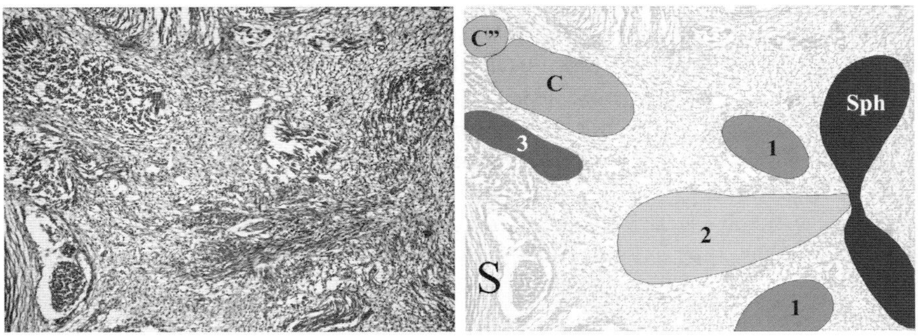

Fig. 20 Section at the level of the ligamentum anococcygeum. On the left side the os sacrum (dorsal), on the right the external sphincter (ventral). *C"*, ligamentum sacrospinale; *C*, m. coccygeus; *S*, os sacrum; *Sph*, m. sphincter ani externus; *1*, m. levator ani; *2*, ligamentum anococcygeum; *3*, m. sacrococcygeus ventralis. Note that the m. levator ani is not yet attached to the m. sphincter ani externus

the midline, following the dorsal part of the m. sphincter ani. Midline attachment of the m. sphincter ani to the left and right m. levator ani is formed by a dense structure of connective tissue, running towards the os coccygis. This structure is the ligamentum anococcygeum. (Fig. 20). The ligamentum anococcygeum is broader than it is in the 50-mm fetus. Therefore midline closure of the pelvis is less advanced.

2.2.4
Musculus Coccygeus

At the surface of the most dorsal part of the corpus ossis ischii (the developing spina ischiadica), the m. coccygeus originates. At this point, the m. levator ani inserts onto the developing spina ischiadica by a band of connective tissue. This is part of the arcus tendineus levatoris ani.

The m. coccygeus rapidly increases in volume as it progresses towards the os coccygis, becoming bigger than the m. levator ani. Just after its origin it becomes clear that the m. coccygeus consists of two parts: a ventral and a dorsal part. This is essentially the same as in the 50-mm fetus, although the separation is less defined. The dorsal part extends towards the lateral surface of the os sacrum, eventually inserting on its surface. This dorsal part of the m. coccygeus is the future ligamentum sacrospinale (Fig. 20). The plexus ischiadicus can be seen running alongside the dorsal edge of the ligamentum sacrospinale.

The ventral part, or actual m. coccygeus, travels towards the midline and continues towards the lateral aspect of the os coccygis. The m. coccygeus is eventually lost in connective tissue before insertion on the surface of the os coccygis and os sacrum. The m. coccygeus can be traced farther dorsally than the m. levator ani and is always separated from the m. levator ani by loose connective tissue. This border between the m. coccygeus and levatoris ani appears to be broader than in the 50-mm fetus.

2.2.5
Musculus Sacrococcygeus Ventralis

The m. sacrococcygeus ventralis is identified as a muscle lying medially from the m. coccygeus (Fig. 20). It originates from the pelvic surface of the os sacrum and inserts on the lateral aspect of the os coccygis. See the description of the 50-mm fetus for further details on the m. sacrococcygeus.

2.2.6
Three-Dimensional Reconstruction Oblique View from Above on the Pelvic Diaphragm

Figure 21 is taken from above, showing the (right) pelvic diaphragm from the inside of the pelvis. In the lower middle is the os sacrum. In the upper left corner the symphysis pubica.

The left and right corpus ossis pubis meet in the midline and form the symphysis pubica (*2*). The rectum and sinus urogenitalis, together forming the corpus pelvinum, have been left out.

The reconstruction of the m. obturatorius internus is shown in red. The ventral to dorsal extension from corpus ossis pubis to the corpus ossis ischii becomes visible. The dorsal extension of the m. obturatorius internus, however, is not as far as in the 50-mm fetus. Dorsally it does not completely cover the ramus and corpus ossis ischii. The m. obturatorius internus also cannot be seen leaving the pelvis to

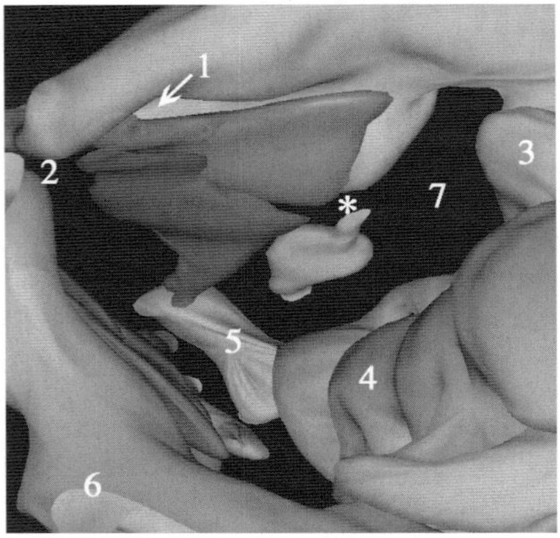

Fig. 21 View from above on the pelvic diaphragm. *1*, canalis obturatorius; *2*, symphysis pubica; *3*, articulatio sacroiliaca; *4*, os sacrum; *5*, os coccygis; *6*, os ilium; *7*, foramen ischiadicum majus; *, future spina ischiadica; *red*, m. obturatorius internus; *blue*, m. levator ani; *green*, m. coccygeus. For illustrative purposes, the pelvic viscera (sinus urogenitalis and rectum) have been left out. The m. levator ani closes the ventral half of the pelvic outlet and almost completely covers the m. obturatorius internus. Compared to the 50-mm fetus, the m. levator ani does not extend as far dorsally. The m. coccygeus closes the dorsal half of the pelvic outlet (full color version of this illustration at the end of the book)

insert on the head of the femur. On the ventral and cranial side the m. obturatorius does not completely reach the lower margin of the ramus superior ossis pubis, as it does in the 50-mm fetus. A completely developed canalis obturatorius is not yet present (*1*). When comparing the extension of the m. obturatorius internus to the 50-mm fetus, it is clear that especially the development in the cranioventral direction is less in the 43-mm fetus.

The reconstruction of the m. levator ani is shown in dark blue. It covers the m. obturatorius internus almost completely, but the dorsal extension is less than in the 50-mm fetus. The m. levator ani in this 43-mm fetus does not reach the level of the future spina ischiadica. The oblique border between the levator ani and the obturatorius internus is in fact the ATLA, running from the inner aspect of the ramus inferior ossis pubis to the later formed spina ischiadica. The m. levator ani has not yet developed far enough dorsally to reach the spina ischiadica. The ATLA already reaches the region of the future spina ischiadica, as can be distinguished in the available slides. This reconstruction of the m. levator ani also makes it distinctly clear that the levator ani consists of a single muscle that covers the frontal half of the pelvic outlet and has a somewhat triangular shape.

The m. coccygeus is reconstructed in green. Its origin is at the level of the later formed spina ischiadica and its insertion is on the surface of the last sacral and first coccygeal vertebra. The m. coccygeus covers the dorsal half of the pelvic outlet. When comparing the m. coccygeus to the 50-mm fetus, it is clear that the muscle is less developed in the 43-mm fetus. Its ventral to dorsal extension is less. The space between the m. levator ani and the surface of the os sacrum is broader. The m. levator ani and the m. coccygeus are two separate muscles, distinguishable from each other by a space between them filled with loose connective tissue. Together they close the pelvic outlet.

The m. sacrococcygeus ventralis is not reconstructed in this 43-mm fetus. It does not cover any part of the actual pelvic outlet and is therefore not considered to belong to the pelvic diaphragm. See the reconstruction of the 50-mm fetus for an example.

2.2.7 Three-Dimensional Reconstruction Frontal and Caudal View of the Pelvic Diaphragm

Figure 22 shows a complete frontal picture of the reconstruction. It is clear that the pelvic outlet of this 43-mm fetus is also projected in the frontal plain, thus pointing forwards. This is in striking contrast to the more backward-oriented pelvic outlet in adult humans. Formation of the angle at the level of the promontorium and the disappearing of the ventral angle between os sacrum and os coccygis is responsible for this backward shifting of the pelvic outlet. The sharp angle of the os coccygis

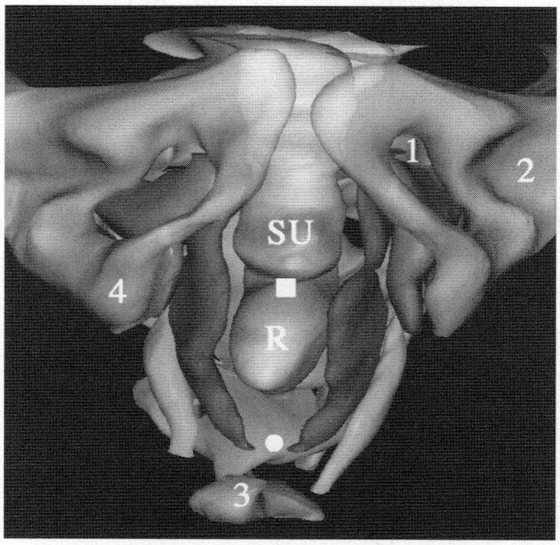

Fig. 22 Frontal view of the pelvic diaphragm. *R*, rectum; *SU*, sinus urogenitalis; *1*, canalis obturatorius; *2*, acetabulum; *3*, os coccygis; *4*, ramus ossis ischii; *white square*, future centrum tendineum; *white dot*, future ligamentum anococcygeum; *red*, m. obturatorius internus; *blue*, m. levator ani; *green*, m. coccygeus. The m. obturatorius internus does not completely cover the foramen obturatum. It is not present near the canalis obturatorius. The left and right mm. levatores ani show curving around the sinus urogenitalis and rectum at the level of the future centrum tendineum and ligamentum anococcygeum. Extension towards the midline is not as advanced as it is in the 50-mm fetus (full color version of this illustration at the end of the book)

to the os sacrum in this fetus is clearly visible in Fig. 23. The forward projection of the pelvic diaphragm in this 43-mm fetus is identical to the 50-mm fetus.

The m. obturatorius internus is reconstructed in red. It covers the foramen obturatum except for a small cranial part. In Fig. 22, it can clearly be seen that the m. obturatorius internus does not reach the lower margin of the ramus superior ossis pubis. A clearly defined canalis obturatorius in this 43-mm fetus is not present as it is in the 50-mm fetus.

Again the m. levator ani is shown in dark blue. It runs from the corpus ossis pubis and lower margin of the ramus inferior ossis pubis to the os coccygis. The left and right m. levator ani surround the corpus pelvinum, which is reconstructed in yellow.

The left and right levator ani are each separate muscles and are curved inward and towards each other at a level between the sinus urogenitalis and the rectum and between the rectum and the os coccygis. The point between the sinus urogenitalis and the rectum will later form the centrum tendineum as the developing left and right levator ani will eventually meet in the midline. The actual formation of a centrum tendineum can be seen in the 50-mm fetus but not in this 43-mm fetus. Behind the rectum, in front of the os coccygis, the left and right levator ani will eventually meet at the level of the developing ligamentum anococcygeum. The distance between the left and right m. levator ani at this point is broader than in the 50-mm fetus. Thus closure of the pelvic outlet in the midline by the two halves of the pelvic diaphragm will eventually take place at the level of the

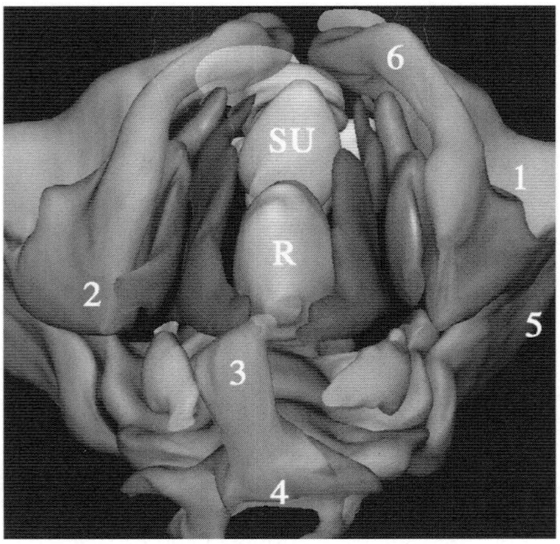

Fig. 23 View of the pelvic diaphragm from below. *R*, rectum; *SU*, sinus urogenitalis; *1*, acetabulum; *2*, ramus ossis ischii/tuber ischiadicum; *3*, os coccygis; *4*, last sacral vertebra; *5*, os ilium; *6*, ramus inferior ossis pubis; *red*, m. obturatorius internus; *blue*, m. levator ani; *green*, m. coccygeus. Note the four-quadrant formation of the pelvic diaphragm. The m. coccygeus is less developed than in the 50-mm fetus (full color version of this illustration at the end of the book)

centrum tendineum and the ligamentum anococcygeum. So closure of the pelvic diaphragm in the midline is not yet present in this fetal stage. Compared to the 50-mm fetus, closure of the pelvic diaphragm in the midline is less advanced in the 43-mm fetus. The insertion of the m. levator ani onto the surface of the os coccygis is not yet completed. A small gap is still present, formed by connective tissue that is not reconstructed.

The m. coccygeus, reconstructed in green, appears to be just as thick as the m. levator ani. In the 50-mm fetus, the m. coccygeus is much thicker. The position of the m. coccygeus is more lateral in relation to the m. levator ani. The two muscles show what looks like a roof tile formation. This is the same as in the 50-mm fetus.

A close look at Fig. 23 suggests that the pelvic diaphragm consists of four quadrants (Marani 2002), as it does in the 50-mm fetus. The ventral left and ventral right quadrants are filled by the left and right m. levator ani and the dorsal quadrants by the m. coccygeus respectively. However, compared to the 50-mm fetus, the covering of the dorsal quadrants by the m. coccygeus is significantly less.

2.3 34 mm (WK 1438) and 35 mm (WK 7608), ± 9 weeks

2.3.1 General Features

These fetuses with a CRL of 34 mm and 35 mm, respectively, are approximately 9 weeks postconception. They are cut transverse in 10-μm sections and stained with HE (see Table 2).

From the appearance of the internal pelvic organs, it is apparent that both these fetuses are female. A developing uterus (fusion of paramesonephric ducts and developing cervix) and two developing ovaries can be clearly identified in the sections through the pelvic organs. Figure 24 shows an example of one of the sections through the pelvic organs in the 34-mm fetus. The section taken is just above the ramus superior ossis pubis and mediolateral oblique. Due to this oblique orientation the right side of the section is just above the ramus superior ossis pubis. The left side of the section is at the level of the acetabulum. On the right side, the vena iliaca externa continuing in the vena femoralis can be seen as the vena femoralis is passing over the ramus superior ossis pubis. Lateral to these vessels lays the m. psoas major and the nervus femoralis dorsal of it. The allantois (future bladder) can be identified, as well as the uterus and the colon sigmoideum, forming part of the corpus pelvinum. On the right side, passing the future uterus, the right ureter can be identified. The corpus pelvinum is embedded in connective tissue with mesenchyme and blood vessels. The density of the connective tissue appears to be less than in the 43-mm fetus. There also seems to be less fat present than in the 43-mm fetus. The uterovesical and rectouterine pouch are visible.

The development of the bony pelvis is nearly complete. The separately developing bones are all outlined in cartilage and no ossification can be noted. At this stage the corpus ossis ilii, the corpus ossis ischii, and the corpus ossis pubis can

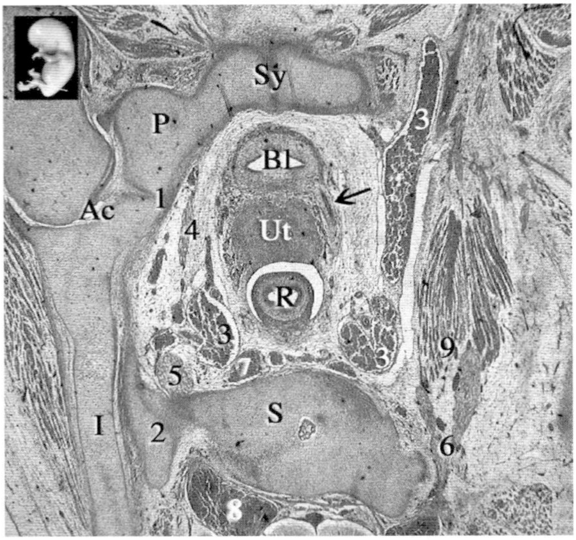

Fig. 24 *Inset*: level of section indicated by a *white line*. *Ac*, acetabulum; *Bl*, allantois (bladder); *I*, os ilium; *P*, os pubis; *R*, rectum/colon sigmoideum; *S*, os sacrum; *Sy*, symphysis pubica; *Ut*, uterus (cervix); *1*, separation os ischium/pubis; *2*, pars lateralis os sacrum; *3*, a.v. iliaca externa/femoralis; *4*, nervus obturatorius; *5*, truncus lumbalis; *6*, nervus femoralis; *7*, presacral lymph node; *8*, spinal cord root; *9*, m. psoas major; *black arrow*, right ureter. As in the 43-mm fetus, the os ilium is surrounded by a thick layer of tissue as a sign of developing cartilage

be identified individually. The borders between the separate bones are more defined than in the 43-mm fetus. Comparable to the 43-mm fetus, several centers of developing cartilage can be seen and the volume of the completed cartilage seems to be less than in the 50-mm fetus. This is especially striking at the level between the os ilium and the os ischii and at the level of the developing labrum glenoidale. A thick layer of tissue surrounds the developing ala ossis ilii and corpus ossis ilii, building up the still thin os ilium. The tuber ischiadicum and spina ischiadica of the os ischii are present as mesenchymal condensations and not as cartilaginous tissue. The ala ossis ilii shows a cartilaginous development centrally. The crista iliaca superiorly, the tuberositas iliaca dorsally and the spina iliaca anterior superior ventrally cannot be distinguished.

The os sacrum of the fetuses is fully developed in cartilage. Fusion of the vertebrae is not yet present. Connection of the os sacrum to the os coxae is formed through ligaments attaching the pars lateralis of the os sacrum to the os coxae. These ligg. sacroiliaca interossea can be identified as such but seem to be far less developed than in the 43-mm and 50-mm fetuses. The os coccygis is formed in cartilage and shows fusion of the third to fifth vertebra.

After evaluating the reconstruction of the cartilaginous pelvis of the 34-mm fetus, it is apparent, like the 50-mm and 43-mm fetuses, that the typical angle at the level of the promontorium is not visible and that the os coccygis makes a sharp angle ventrally. Absence of the angle at the level of the promontorium and the sharp angle of the os coccygis seem responsible for the ventral position of the pelvic outlet present in this fetus. This is in accordance with the older fetuses.

2.3.2 Musculus Obturatorius Internus

At this stage, the m. obturatorius internus itself does not completely cover the foramen obturatum. Especially at the level of the os pubis the muscle fibers do not reach the surface of the inner aspect of the corpus ossis pubis, i.e., the ramus superior and inferior ossis pubis. The fascia of the m. obturatorius internus, however, is attached to the corpus ossis pubis. On the dorsal side, the m. obturatorius internus reaches the most dorsal part of the ramus ossis ischii only at a level just above the developing tuber ischiadicum. The muscle itself cannot be seen leaving the pelvic cavity as it does in the 50-mm fetus, where it can be seen to insert on the femur. In fact, at this 34/35-mm stage, only the fascia of the m. obturatorius internus extends to the trochanter major femoris where it inserts.

At its lateral side, the m. obturatorius is covered by the membrana obturatoria. The membrana obturatoria is about as thick as in the 43-mm fetus but thicker than in the 50-mm fetus. On the other side of the membrana obturatoria, the m. obturatorius externus can be identified (Fig. 25).

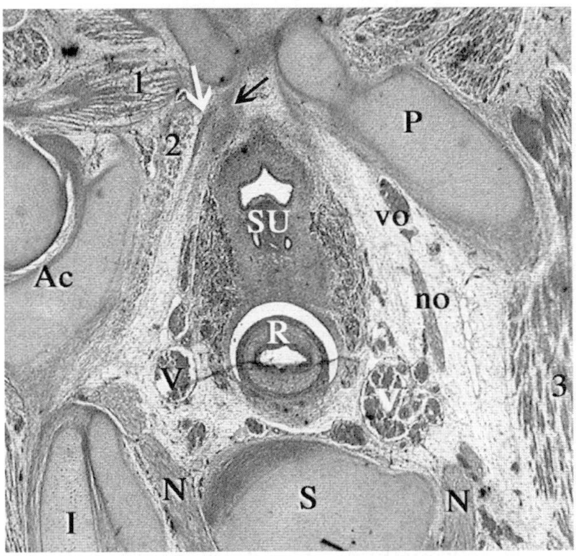

Fig. 25 *Ac*, acetabulum; *I*, os ilium; *N*, nervus sacralis I; *no*, nervus obturatorius; *P*, ramus superior ossis pubis; *R*, rectum; *S*, os sacrum; *SU*, sinus urogenitalis; *V*, vena iliaca interna; *vo*, vena obturatoria; *1*, m. obturatorius externus; *2*, m. obturatorius internus; *3*, m. psoas major; *white arrow*, arcus tendineus levatoris ani; *black arrow*, endopelvic fascia

Following the sections, the nervus obturatorius can be seen leaving the pelvis just below the ramus superior ossis pubis. The nerve, however, is not completely surrounded by the m. obturatorius internus. The muscle starts to appear well below the level of the nervus obturatorius and is not present on the ventral and dorsal side of the nervus obturatorius. The cranial extension of the m. obturatorius internus is less than in the 43-mm fetus. This means that the foramen obturatum is not covered by the m. obturatorius internus on the cranial side. The canalis obturatorius as such is not yet constituted.

Just below the level of the symphysis, on the inner aspect of the ramus inferior ossis pubis, a broad band of connective tissue can be seen. It runs towards the sinus urogenitalis and eventually radiating around it. This is the developing pubovesical ligament (considered as the ventral part of the ATLA in adult human anatomy) and the endopelvic fascia. Compared to the later stages, the ligament and fascia are less condensed. This is considered as a lesser development of the ligament and the endopelvic fascia in the 34-mm fetus than in the 43-mm and 50-mm fetuses.

In between the developing endopelvic fascia and the m. obturatorius internus, a thin and dense band of connective tissue appears. It appears at a level that is about halfway the foramen obturatum and well below the nervus obturatorius. This band of connective tissue is the arcus tendineus levatoris ani (or arcus tendineus fascia pelvis). The ATLA could be identified in the 43-mm fetus as a broad band of connective tissue. Compared to the 50-mm fetus, the ATLA is less developed. It does not reach the inner aspect of the ramus inferior ossis pubis ventrally. Dorsally it does not extend as far as in the 50-mm fetus. When it arises, it covers about a quarter of the medial aspect of the m. obturatorius. It fades away in loose connective tissue that extends to about halfway the m. obturatorius internus (Fig. 25).

2.3.3 Musculus Levator Ani

Following the ATLA through the fetus, it starts to show a gradual thickening on the ventral side as muscle fibers start to appear. The muscle fibers are in a straight angle to the ATLA. When it originates, the m. levator ani does not show such a close relationship with the inner aspect of the ramus inferior ossis pubis as it does in the 50-mm fetus. Its ventral margin of muscle fibers is farther away from the os pubis. The gap is filled by undefined connective tissue that will probably play a role in the development of the diaphragma urogenitalis. At its origin, the dorsal extension of the m. levator ani is over about a quarter of the medial aspect of the m. obturatorius internus. In the 50-mm fetus, it immediately covered more than half the medial aspect and in the 43-mm fetus about half.

The moment it originates, it is again clear that the m. levator ani is one single muscle. It stays separated from the m. obturatorius internus by a small margin. As in the 43-mm fetus, this close relationship is maintained all the way to the ramus ossis ischii at the point where the m. obturatorius internus leaves the pelvis.

Progressing caudally, the thickness of the m. levator ani starts to increase, especially on its ventral side. The ventral to dorsal length increases and at the upper level of the ramus ossis ischii the m. levator ani has developed to a muscle covering the m. obturatorius internus for about half its length. The m. levator ani is at this point not yet fixed to the dorsal end of the corpus ossis ischii. The decreasing ventral–dorsal length of the m. obturatorius internus mostly contributes to this partial covering. The covering of the m. obturatorius internus at this stage is not as complete as it is in the 43-mm fetus. Especially on the dorsal side, near the corpus ossis ischii, the m. levator ani does not completely cover the m. obturatorius internus. Complete covering starts to appear near the level of the developing tuber ischiadicum where the m. levator ani attaches to the surface of the developing spina ischiadica. At this level, the m. obturatorius internus completely disappears. At the point where the m. levator ani attaches to the developing spina ischiadica a third muscle originates. This is the m. coccygeus.

Passing around the sinus urogenitalis, the ventral side of the m. levator ani is attached to connective tissue that appears to be the developing diaphragma urogenitalis. The m. levator ani is nowhere directly attached to the sinus urogenitalis, but only lies closely adjacent to it. Behind the sinus urogenitalis, no connection can be identified between the left and right m. levator ani. They are separated from each other by connective tissue that does not have the appearance of a centrum tendineum like it has in the 50-mm fetus. In between the sinus urogenitalis and the rectum the m. sphincter ani externus starts to appear. The m. levator ani is not attached to the m. sphincter ani externus, nor is it attached to the rectum (Fig. 26). Behind the rectum, the left and right m. levator ani start to run towards the midline. Eventually the left and right m. levator ani are connected to each other and to the most dorsal part of the m. sphincter ani externus by connective tissue. This connective tissue can be followed from the deep part of the m. sphincter ani externus to the tip of the os coccygis. It is the developing ligamentum anococcygeum (Fig. 27), which shows about the same grade of development as it does in the 43-mm fetus.

Compared to the 43-mm fetus, the m. levator ani in this 34-mm fetus shows the same position in relation to the corpus pelvinum. But the m. levator ani in the 34-mm fetus does not extend as much dorsally as it does in the 43-mm and 50-mm fetus.

2.3.4 Musculus Coccygeus

At the surface of the most dorsal part of the corpus ossis ischii, i.e., the future spina ischiadica, the m. coccygeus originates (Fig. 26). The m. levator ani also attaches to the surface of the os ischii at this point. Attachment is made by connective tissue, that is the arcus tendineus levatoris ani. In contrast to the 43-mm fetus, the ATLA as such cannot be identified at the level of the os ischii and developing spina ischiadica.

After its origin, the m. coccygeus increases in volume as it progresses towards the os coccygis but slower when compared to the 43-mm fetus. As in the later stages, the m. coccygeus soon appears to consist of two parts (Fig. 26). In this 34-mm fetus a branch of the nervus pudendus can be seen passing in between the ventral and dorsal part. The dorsal part, that seems to be relatively bigger in this fetus when compared to the older fetuses, spreads out towards the surface of the os sacrum. This dorsal part is the future ligamentum sacrospinale.

The ventral part, or actual m. coccygeus, travels towards the midline. It can be followed up to its insertion on the lateral aspect of the last coccygeal vertebrae (Fig. 28). Its relationship with the ligamentum anococcygeum is not clear.

The m. coccygeus has no relationship or contact with the m. levator ani. The m. coccygeus runs dorsal and lateral from the m. levator ani and is separated by a border of loose connective tissue. This border appears to be broader than in the 43-mm fetus. The m. coccygeus can also be followed farther towards the os coccygis than the m. levator ani. The levator ani inserts on to the os coccygis via the ligamentum anococcygeum.

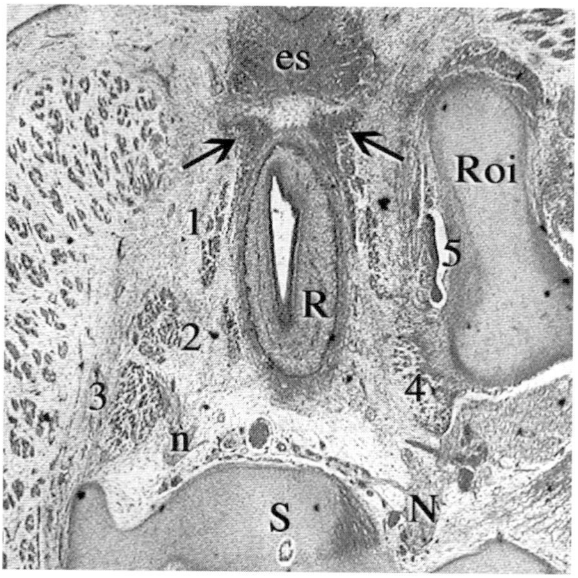

Fig. 26 *es*, m. sphincter ani externus; *N*, nervus pudendus; *n*, branch of n. pudendus; *R*, rectum; *Roi*, ramus ossis ischii; *S*, os sacrum; *black arrow*, m. sphincter ani externus (deep part); *1*, m. levator ani; *2*, ventral part m. coccygeus; *3*, dorsal part m. coccygeus (ligamentum sacrospinale); *4*, origin m. coccygeus at developing spina ischiadica. The deep part of the m. sphincter ani externus extends around the rectum, attaching to the m. levator ani. A branch of the nervus pudendus can be seen piercing between the ventral and dorsal part of the m. coccygeus

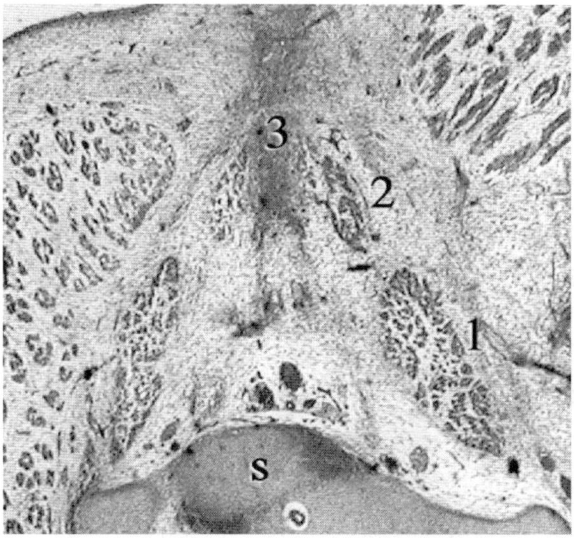

Fig. 27 *S*, os sacrum; *1*, m. coccygeus; *2*, m. levator ani; *3*, ligamentum anococcygeum. The m. levator ani on the left side is fixed to the anococcygeal ligament. This ligament runs between the m. sphincter ani externus and os coccygeus

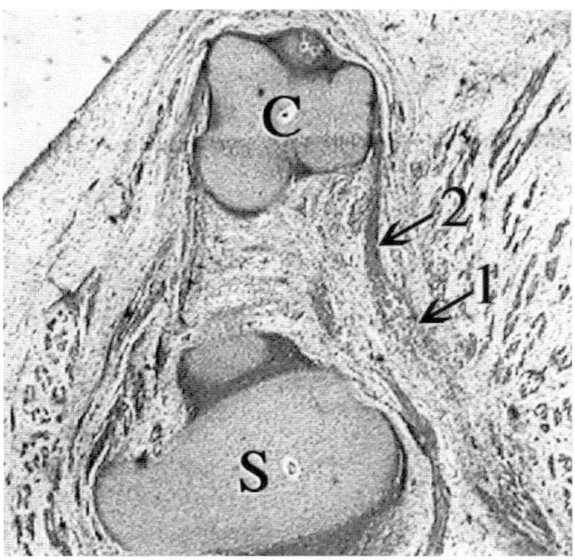

Fig. 28 *C*, os coccygis; *S*, os sacrum; *1*, m. coccygeus; *2*, fascia of the m. coccygeus inserting on lateral aspect of os coccygis

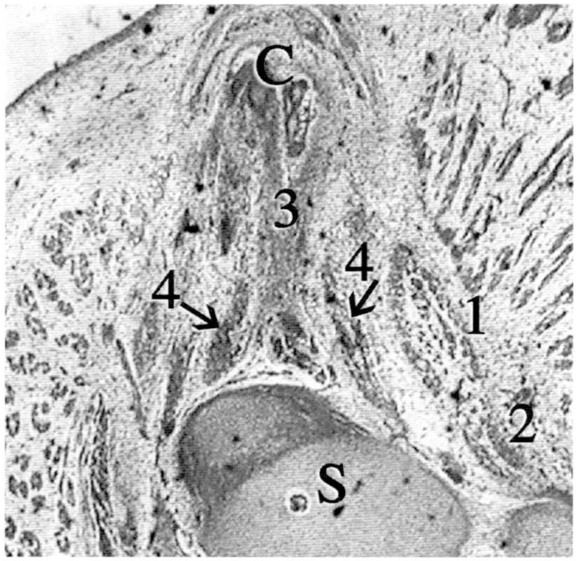

Fig. 29 *C*, tip of os coccygis (last vertebra); *S*, os sacrum; *1*, ventral part of m. coccygeus; *2*, dorsal part of m. coccygeus (ligamentum sacrospinale); *3*, ligamentum sacrococcygeum ventrale; *4*, m. sacrococcygeus ventralis

2.3.5 Musculus Sacrococcygeus Ventralis

The m. sacrococcygeus ventralis is identified as a muscle lying medially from the m. coccygeus (Fig. 29). It originates from the pelvic surface of the os sacrum and inserts on the lateral aspect of the os coccygis. One should look into the description of the 50-mm fetus for further details on the m. sacrococcygeus.

2.3.6 Three-Dimension Reconstruction Oblique View from Above on the Pelvic Diaphragm

This picture of the three-dimensional reconstruction (Fig. 30) is taken from above, showing the (right) developing pelvic diaphragm from the inside of the pelvis. In the lower right corner is the os sacrum. In the upper left corner the symphysis pubica can be noted.

The left and right corpus ossis pubis meet in the midline and are forming the symphysis pubica (*2*). The rectum and sinus urogenitalis, together forming part of the corpus pelvinum, have been left out in this reconstruction to give a better view on the muscles of the pelvic diaphragm.

The reconstruction of the m. obturatorius internus is shown in red. The ventral to dorsal extension from corpus ossis pubis to the corpus ossis ischii is visible. The dorsal extension of the m. obturatorius internus is about as far as in the 50-mm fetus. Dorsally, it does not completely cover the ramus and corpus ossis ischii. The m.

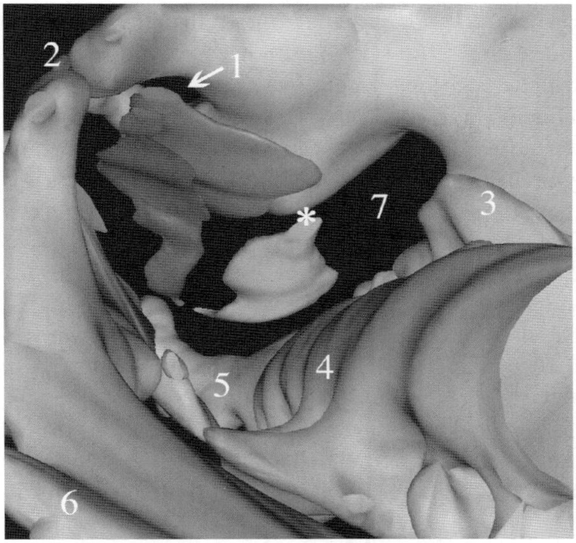

Fig. 30 View from above on the pelvic diaphragm. *1*, foramen obturatum; *2*, symphysis pubica; *3*, pars lateralis os sacrum; *4*, os sacrum; *5*, os coccygis; *6*, ala ossis ilii; *7*, foramen ischiadicum majus; *, future spina ischiadica; *red*, m. obturatorius internus; *blue*, m. levator ani; *green*, m. coccygeus (full color version of this illustration at the end of the book)

obturatorius internus itself can also not be seen leaving the pelvis to insert on the head of the femur. However, the fascia of the muscle does, but it is not reconstructed. On the ventral and cranial side, the m. obturatorius does not reach the lower margin of the ramus superior ossis pubis. Covering of the upper half of the foramen obturatum is not present. In the 43-mm embryo, the m. obturatorius internus fills in approximately three-quarters of the foramen obturatum, since it is absent at the cranial part. A canalis obturatorius as such is not present (*1*). When comparing the extension of the m. obturatorius internus to the 43-mm fetus, it is clear that the development in the dorsal direction is roughly the same as in the 43-mm fetus. The development in the ventral and cranial direction is less compared to the 43-mm fetus.

The reconstruction of the m. levator ani is shown in dark blue. Near the symphysis pubica, roughly one-quarter of the m. obturatorius internus is covered. More distally, about half of the m. levator ani covers the m. obturatorius internus. The m. levator ani in this 34-mm fetus does not reach the level of the future spina ischiadica. The oblique border between the levator ani and the obturatorius internus near its origin at the level of the symphysis pubica is in fact the ATLA, running from the inner aspect of the ramus inferior ossis pubis to the future developing spina ischiadica. The m. levator ani has not yet developed far enough dorsally to reach most of the ATLA. Whether the ATLA already reaches the region of the future spina ischiadica cannot be distinguished in the available slides. This reconstruction of the m. levator ani also makes it distinctly clear that the levator ani consist of a single muscle that covers the frontal half of the pelvic outlet. In contrast to the 43-mm and 50-mm fetuses, it does not yet have the typical triangular shape.

The m. coccygeus is reconstructed in green. Its origin is at the level of the future spina ischiadica and its insertion is on the surface of the last sacral and first coccygeal vertebrae. The m. coccygeus covers the dorsal half of the fetal pelvic outlet. When comparing the m. coccygeus to the 43-mm fetus, it is clear that the muscle shows about the same development. The dorsal extension is farther towards the os sacrum than observed in the 43-mm fetus. Moreover, the space between the m. levator ani and the surface of the os sacrum is broader. This is due to the fact that the dorsal extension of the m. levator ani in the 34-mm fetus is less than in the 43-mm fetus. The m. levator ani and the m. coccygeus are two separate muscles, distinguishable from each other by a space filled with connective tissue. Together they close the pelvic outlet.

The m. sacrococcygeus ventralis is not reconstructed in this 34-mm fetus. It does not cover any part of the actual pelvic outlet and is therefore not considered to belong to the pelvic diaphragm (see the reconstruction of the 50-mm fetus for an example).

2.3.7 Three-Dimensional Reconstruction Frontal and Caudal View of the Pelvic Diaphragm

Figure 31 shows a complete frontal picture of the reconstruction. It is clear that the pelvic outlet of this 43-mm fetus is also projected in the frontal plain, thus pointing forwards. This is in striking contrast to the more backward-oriented pelvic outlet in adult humans. Formation of the angle at the level of the promontorium and the disappearing of the ventral angle between os sacrum and os coccygis is responsible for this backward shifting of the pelvic outlet present after birth. The sharp angle of the os coccygis to the os sacrum in this fetus is clearly visible in Fig. 32. The forward projection of the pelvic diaphragm in this 34-mm fetus is identical to the 43-mm and 50-mm fetuses.

The m. obturatorius internus is reconstructed in red. It covers the lower three-quarters of the foramen obturatum. In Fig. 31, it is clearly visible that the m. obturatorius internus does only cover the lower part of the foramen obturatum. A clearly defined canalis obturatorius in this 34-mm fetus is not present, as it is in the 50-mm fetus.

Again the m. levator ani is shown in dark blue. It runs from the corpus ossis pubis and lower margin of the ramus inferior ossis pubis to the os coccygis. The left and right m. levator ani surround the corpus pelvinum, which is reconstructed in yellow.

The left and right levator ani are each separate muscles and are curved inwards and towards each other at a level between the sinus urogenitalis and the rectum and between the rectum and the os coccygis. This curving in the 34-mm fetus, however, is less pronounced than in the 43-mm en 50-mm fetuses. The area between the sinus urogenitalis and the rectum will form the centrum tendineum

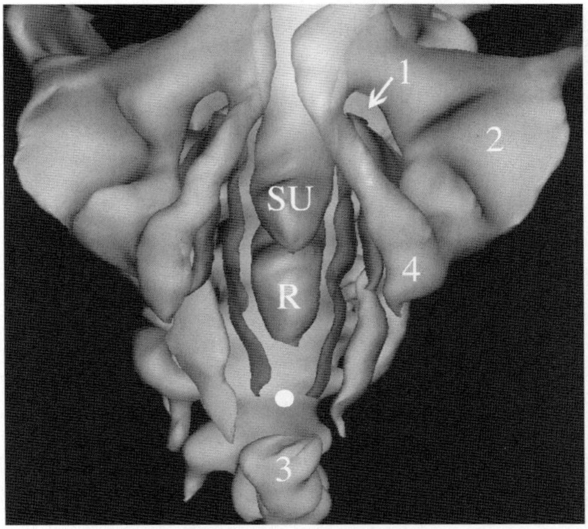

Fig. 31 Frontal view of the pelvic diaphragm. *R*, rectum; *SU*, sinus urogenitalis; *1*, foramen obturatum; *2*, acetabulum; *3*, os coccygis; *4*, ramus ossis ischii; *white dot*, future ligamentum anococcygeum; *red*, m. obturatorius internus; *blue*, m. levator ani; *green*, m. coccygeus (full color version of this illustration at the end of the book)

as the developing left and right levator ani will eventually reach the centrum tendineum. The actual developing centrum tendineum is present in the 50-mm fetus but not in the 34-mm fetus. Behind the rectum, in front of the os coccygis, the left and right levator ani will eventually reach the developing ligamentum anococcygeum. The distance between the left and right m. levator ani at this point is less than in the 43-mm fetus. Closure of the pelvic diaphragm in the midline is not present in this fetal stage. Compared to the 43-mm fetus, closure of the pelvic diaphragm in the midline is less advanced in this 34-mm fetus. The insertion of the m. levator ani onto the surface of the os coccygis is also not yet completed. A small gap is still present, formed by connective tissue that is not reconstructed.

The m. coccygeus, reconstructed in green, appears to be slightly thicker than the m. levator ani. In the 50-mm fetus, the m. coccygeus is in fact much larger. The position of the m. coccygeus is more lateral in relation to the m. levator ani. The two muscles show what looks like a roof tile formation. This is the same spatial position as in the 43-mm and 50-mm fetuses.

A close look at Fig. 32 suggests that the pelvic diaphragm consists of four quadrants, as it does in the older fetuses. The ventral left and ventral right quadrants are filled by the left and right m. levator ani and the dorsal quadrants by the m. coccygeus. Compared to the 43-mm fetus, the covering of the dorsal quadrants by the m. coccygeus is slightly more dorsally directed in this 34-mm fetus. The covering of the ventral quadrants by the m. levator ani, however, is clearly less advanced in the 34-mm fetus.

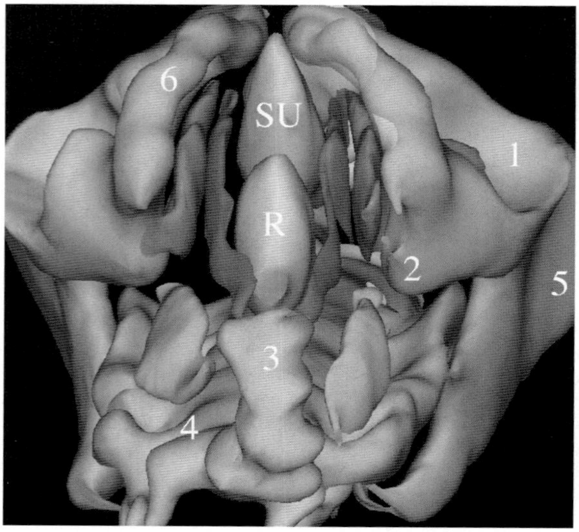

Fig. 32 View of the pelvic diaphragm from below. *R*, rectum; *SU*, sinus urogenitalis; *1*, acetabulum; *2*, ramus ossis ischii at level of developing spina ischiadica; *3*, os coccygis; *4*, os sacrum; *5*, os ilium (ala ossis ilii); *6*, ramus inferior ossis pubis; *red*, m. obturatorius internus; *blue*, m. levator ani; *green*, m. coccygeus (full color version of this illustration at the end of the book)

3 Embryonal Period

3.1 Stage 23: 28 mm (WK 1451) and 30 mm (WK 7652), ± 58–59 days

3.1.1 General Features

These embryos with a CRL of 28 mm and 30 mm, respectively, are approximately 8.5 weeks (58–59 days) postconception. The embryos are in their developmental stage 23 according to the subdivision by Streeter (1942–1951, described in Gribnau and Geijsberts 1981). This stage is the end of the embryonic development, thus just before entering the fetal period. This embryological stage is characterized by the complete separation of the fingers and the toes. The embryos are cut transverse in 10-μm sections and stained with HE (see Table 2).

Figure 33 shows a section through the pelvic cavity of the 28-mm embryo. The pelvic cavity of the 30-mm embryo has an identical appearance. In between the allantois (future bladder) and the hindgut (future rectum), the urogenital septum is localized. In the center of this urogenital septum are the mesonephric ducts (Wolffian) laterally and the partially fused paramesonephric ducts (Müllerian) medially. Both embryos are female.

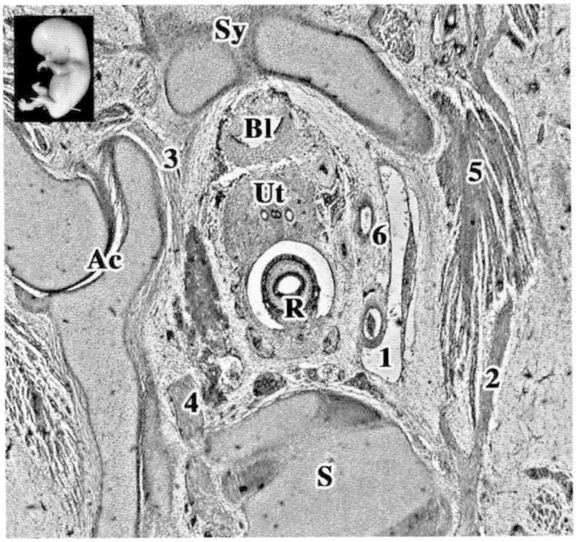

Fig. 33 *Inset*: level of section indicated by a *white line*. *Ac*, acetabulum; *Bl*, allantois (bladder); *R*, hindgut; *S*, last lumbar vertebra; *Sy*, symphysis pubica; *Ut*, uterus (cervix); *1*, a.v. iliaca externa; *2*, nervus femoralis; *3*, nervus obturatorius; *4*, nervus lumbalis V; *5*, m. psoas major; *6*, arteria umbilicalis

As in the 34-mm fetus, the 28-mm embryo is cut obliquely mediolaterally. This orientation is well illustrated in Fig. 33. The left side of the section runs at the level of the foramen obturatum and the acetabulum. The right side of the section is located above the ramus superior ossis pubis at the level of the m. psoas major. On the left side, the nervus obturatorius can be seen leaving the pelvic cavity. At the right side, the arteria umbilicalis is clearly visible. The developing hindgut, the developing uterus/cervix and the allantois (together forming the corpus pelvinum) are embedded in developing connective tissue. The density of the developing connective tissue appears to be the same as in the 34-mm fetus but there is hardly any sign of fat present. The extension of the peritoneal cavity as a rectouterine pouch is visible, but the presence of an uterovesical pouch is less defined as in the 34-mm and older fetuses.

The development of the bony pelvis is not yet complete in these embryos. The bones present are all outlined only in cartilage. No ossification is noted. Nevertheless, the corpus ossis ilii, the corpus ossis ischii and the corpus ossis pubis can be identified individually. The individual development of especially the corpus ossis pubis and corpus ossis ischii is more clearly outlined in developing cartilage compared to the older fetuses. The corpus ossis ilii is completely present but its ala ossis ilii is far less developed than it is in later stages. The tuber ischiadicum of the os ischii is solely present as a mesenchymal condensation. The spina ischiadica, however, is far less outlined in mesenchyme as it is in the 34-mm and older fetuses.

The os sacrum of the embryos is fully developed in cartilage. Fusion of the vertebrae is not yet present. Connections of the os sacrum to the os coxae is formed through mesenchyme connecting the pars lateralis of the os sacrum to the os coxae. The ligg. sacroiliaca interossea, as present in the older fetuses, cannot be identified as such. The os coccygis is formed in cartilage and shows fusion of the third to fifth vertebra.

After evaluating the reconstruction of the cartilaginous pelvis of the 28-mm fetus, it is apparent, like the 50-mm, 43-mm and 34-mm fetuses, that the typical angle at the level of the promontorium is not visible and that the os coccygis makes a far sharper angle ventrally. Absence of the angle at the level of the promontorium and the sharp angle of the os coccygis seems responsible for the ventral position of the pelvic outlet present in this embryonic stage. This is in accordance with the older fetuses.

3.1.2 Musculus Obturatorius Internus

Similar to the later stages, the m. obturatorius internus itself does not completely cover the foramen obturatum. Especially at the level of the os pubis, the muscle fibers do not reach the surface of the inner aspect of the corpus ossis pubis, i.e., the ramus superior and the ramus inferior ossis pubis. The fascia of the m. obturatorius internus, however, appears not to be attached to the corpus ossis pubis yet, but this is difficult to see in the available material. This is in striking contrast with the 34-mm fetus where the fascia of the m. obturatorius internus is already fixed to the inner aspect of the os pubis. This must mean that the ventral part of the foramen obturatum is not covered by the m. obturatorius internus at the end of the embryological period. On the dorsal side, the m. obturatorius internus reaches the most dorsal part of the ramus ossis ischii only at the level just above the developing tuber ischiadicum. The muscle itself cannot be located outside the pelvic cavity as it is in the 50-mm fetus, where it can be seen to insert on the femur. In fact, only the fascia of the m. obturatorius internus extends to the trochanter major femoris where it clearly inserts.

At its lateral side, the m. obturatorius is covered by the membrana obturatoria. The membrana obturatoria in the 28-mm embryo is thicker than it is in the 34-mm fetus. On the other side of the membrana obturatoria, the m. obturatorius externus is clearly developed (Fig. 34).

The nervus obturatorius can be seen leaving the pelvis just below the ramus superior ossis pubis, following the sections. The nerve, however, is free from the m. obturatorius internus. The muscle starts to appear below the level of the nervus obturatorius and is not present at the ventral or at the dorsal side of the nervus obturatorius. The cranial extension of the m. obturatorius internus is about the same as it is in the 34-mm fetus. This means that the foramen obturatum is not covered by the m. obturatorius internus on the cranial side. The canalis obturatorius as such is not yet constituted.

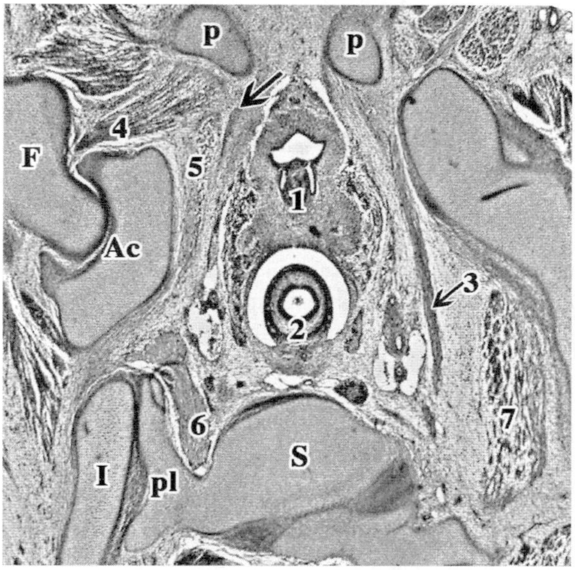

Fig. 34 *Ac*, acetabulum; *F*, femur; *I*, os ilium; *P*, os pubis (ramus inferior); *pl*, pars lateralis os sacrum; *S*, os sacrum; *1*, sinus urogenitalis; *2*, hindgut; *3* (+*arrow*), nervus obturatorius; *4*, m. obturatorius externus; *5*, m. obturatorius internus; *6*, plexus lumbosacralis (ramus ventralis nervus sacralis I); *7*, m. psoas major; *black arrow*, arcus tendineus levatoris ani

Just below the level of the symphysis, on the inner aspect of the ramus inferior ossis pubis, a broad band of connective tissue can be seen. It runs towards the sinus urogenitalis and eventually radiating around it. This is the developing pubovesical ligament (considered as the ventral part of the ATLA in adult human anatomy) and the endopelvic fascia. The development of the ligament and fascia is the same in the 34-mm fetus.

Between the developing endopelvic fascia and the m. obturatorius internus, a thin and dense band of connective tissue appears. It is located at a level that is roughly halfway over the foramen obturatum and well below the nervus obturatorius. This band of connective tissue is the arcus tendineus levatoris ani (ATLA). The ATLA shows a similar development as it does in the 34-mm fetus. It cannot be discerned at the inner aspect of the ramus inferior ossis pubis ventrally. Dorsally, it extends about as far as it does in the 34-mm fetus. When it arises, it covers roughly one-quarter of the medial aspect of the m. obturatorius. It fades away in loose connective tissue that covers the ventral half of the m. obturatorius internus.

3.1.3 Musculus Levator Ani

Following the ATLA through the embryo, the m. levator ani shows a gradual thickening on the ventral side as muscle fibers start to originate. The muscle fibers are in a straight angle to the ATLA. When it originates, the m. levator ani does not show such a close relationship with the inner aspect of the ramus inferior ossis pubis as it does in the older fetuses. The distance between the left and right os pubis at the level of the muscle's origin is bigger than it is in the 34-mm fetus. The ventral margin of the first appearing muscle fibers is also farther away from the os pubis compared to the 34-mm fetus. The gap is filled by, until now, undefined connective tissue that will probably play a role in the development of the diaphragma urogenitalis. At its origin, the dorsal extension of the m. levator ani is over approximately one-quarter of the medial aspect of the m. obturatorius internus. This configuration is also found in the 34-mm fetus. In the 43-mm fetus, it immediately covers roughly half the medial aspect and in the 50-mm fetus more than half.

The moment it originates and can be recognized, it is again clear that the m. levator ani is a single muscle. It stays separated from the m. obturatorius internus by a small margin of developing connective tissue. As in the 34-mm and 43-mm fetuses, this separate, but close relationship is maintained all the way to the lower end of the corpus ossis ischii at the point where the m. obturatorius internus leaves the pelvis. Below this point, above the developing ramus ossis ischii, this relationship is lost as the m. levator ani starts to extend more medially. There appears an increasing gap between the m. levator ani and m. obturatorius internus that is now filled with loose connective tissue. This gap is the fossa ischiorectalis, which is bigger compared to the older fetuses. At the level of insertion of the m. obturatorius internus near the developing tuber ischiadicum, the nervus pudendus starts to appear.

Progressing caudally through the sections the thickness of the m. levator ani increases especially on its ventral side. The ventral to dorsal length increases and at the upper level of the ramus ossis ischii, the m. levator ani has developed to a muscle covering the m. obturatorius internus completely (Fig. 35). A decreasing ventral–dorsal length of the m. obturatorius internus is responsible for this complete covering. The ventral to dorsal extension of the m. levator ani is similar to the 34-mm fetus. At a level just above the developing tuber ischiadicum, the dorsal extension of the m. levator ani surpasses the m. obturatorius internus. At this point, the m. levator ani reaches its maximal ventral to dorsal extension. Apart from its fascia, the m. levator ani is not fixed to the dorsal end of the corpus ossis ischii. Behind the m. levator ani, a separate muscle starts to appear. This is the m. coccygeus.

Passing around the sinus urogenitalis, the ventral side of the m. levator ani is attached to connective tissue that appears to be the developing diaphragma (hiatus) urogenitalis. The m. levator ani is totally free from the sinus urogenitalis. It lies closely adjacent to it and shows some inward curving but stays unrelated to this sinus. Behind the sinus urogenitalis, no connection can be identified between the left and right m. levator ani. They are separated from each other by connective tissue that does not have the appearance of a centrum tendineum like it has in the 50-mm fetus. In these embryos, the relationship of the m. levator ani to the sinus urogenitalis is comparable to the 34-mm and 43-mm fetuses, in which it also stays free.

In between the sinus urogenitalis and the rectum, the m. sphincter ani externus starts to appear in the more caudal sections. The deep part of the m. sphincter

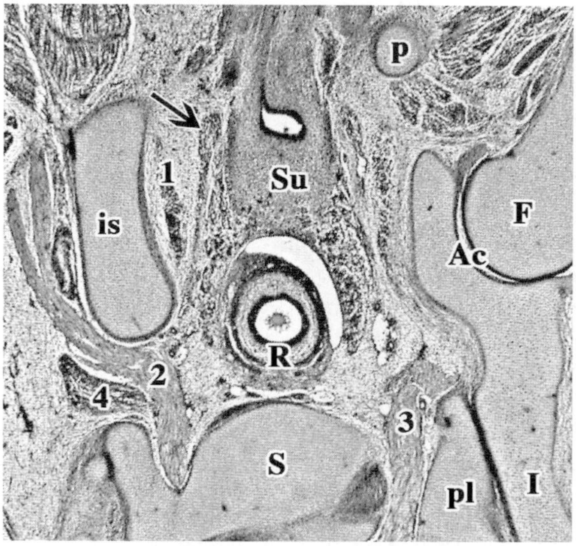

Fig. 35 *Ac*, acetabulum; *F*, femur (caput); *is*, os ischii; *I*, os ilium; *p*, os pubis (ramus inferior); *pl*, pars lateralis os sacrum; *R*, hindgut; *S*, os sacrum; *Su*, sinus urogenitalis; *1*, m. obturatorius internus; *2*, nervus/plexus ischiadicus; *3*, nervus sacralis I (plexus lumbosacralis); *4*, m. piriformis; *black arrow*, m. levator ani

externus seems to have a closer relationship with the m. levator ani compared to the older fetuses. The m. levator ani is attached to the m. sphincter ani externus. It is not attached to the rectum (Fig. 36). Behind the rectum, the left and right m. levator ani start to run towards the midline. Eventually the left and right m. levator ani near each other and the most dorsal part of the m. sphincter ani externus via connective tissue. This connective tissue can be followed from the deep part of the m. sphincter ani externus to the tip of the os coccygis. This is the developing ligamentum anococcygeum (Fig. 37). Compared to the 34-mm fetus, these structures are less defined.

3.1.4 Musculus Coccygeus

At the surface of the most dorsal part of the corpus ossis ischii, i.e., the future spina ischiadica, the m. coccygeus originates (Fig. 36). The fascia of the m. levator ani also attaches to the surface of the os ischii at this point. Attachment is made by connective tissue that is the arcus tendineus levatoris ani. Comparable to the 34-mm fetus, the ATLA as a well defined structure cannot be identified at the level of the os ischii and developing spina ischiadica.

After its origin, the m. coccygeus increases in volume as it progresses towards the os coccygis. The way it increases in volume and the total volume reached is about the same as in the 34-mm fetus. As in the later stages, the m. coccygeus soon

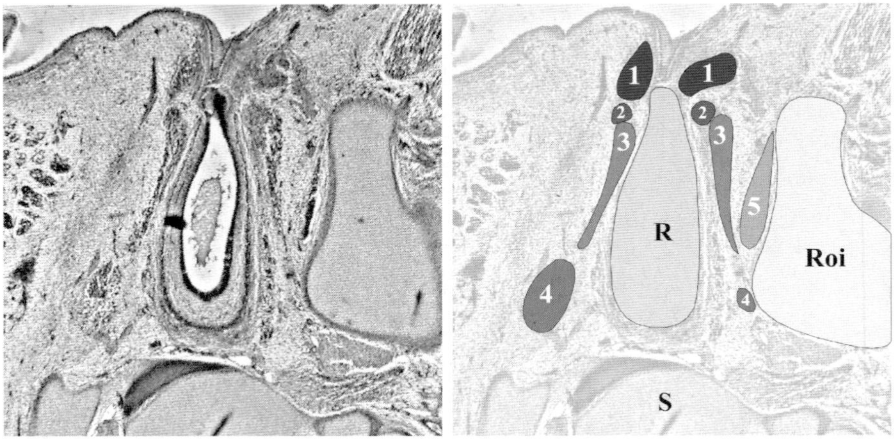

Fig. 36 *R*, hindgut; *Roi*, ramus/corpus ossis ischii; *S*, os sacrum; *1*, superficial part of m. sphincter ani externus; *2*, deep part of m. sphincter ani externus; *3*, m. levator ani; *4*, m. coccygeus; *5*, m. obturatorius internus. Although hardly visible, there is a small border between the superficial and deep part of the external sphincter. There is a close relationship between the deep part of the m. levator ani and the deep part of the external sphincter. They are directly attached to each other

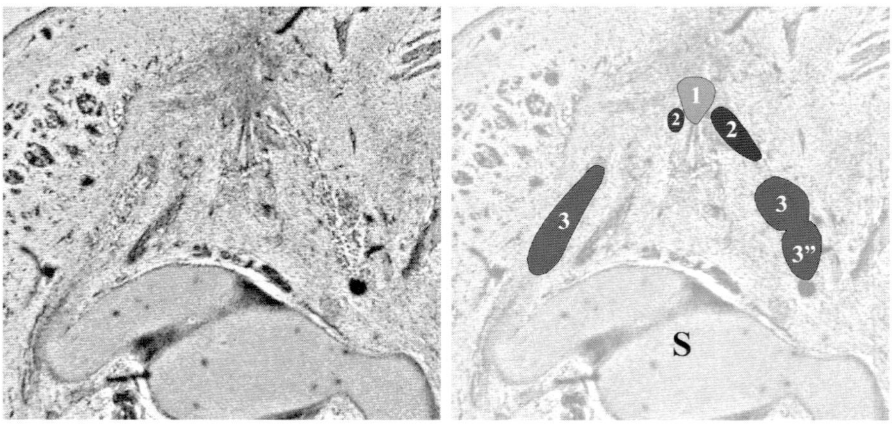

Fig. 37 *S*, os sacrum; *1*, developing ligamentum anococcygeum; *2*, m. levator ani; *3*, m. coccygeus; *3"*, dorsal part of m. coccygeus or future ligamentum sacrospinale. The m. coccygeus on the left side shows the same division in a ventral and dorsal part as seen on the right side, in more rostral sections

appears to consist of two parts (Fig. 37). The division between the two parts is less defined than in the older fetuses. The dorsal part is the same size as the older fetuses and is larger than its ventral part. The dorsal part spreads out towards the surface of the os sacrum and becomes the ligamentum sacrospinale.

The ventral part, or actual m. coccygeus, travels towards the midline. Its fascia can be followed up to its insertion on the lateral aspect of the last coccygeal vertebrae. The m. coccygeus has no relationship with the ligamentum anococcygeum that inserts on the tip of the os coccygis (Fig. 37).

The m. coccygeus has no relationship or contact with the m. levator ani. The m. coccygeus runs dorsal and lateral from the m. levator ani and is separated by a border of loose connective tissue. The extension of this border of connective tissue is the same as in the 34-mm fetus. The m. coccygeus can also be followed farther towards the os coccygis than the m. levator ani. The levator ani inserts on to the os coccygis via the ligamentum anococcygeum.

3.1.5 Musculus Sacrococcygeus Ventralis

The m. sacrococcygeus ventralis is identified as a muscle lying medially from the m. coccygeus. Compared to the older fetuses, it is much smaller and less outlined. It originates from the pelvic surface of the os sacrum and inserts on the lateral aspect of the os coccygis. One should look into the description of the 50-mm fetus for further details on the m. sacrococcygeus.

3.1.6 Three-Dimensional Reconstruction Oblique View from Above on the Pelvic Diaphragm

Figure 38, illustrating the three-dimensional reconstruction, is taken from above and shows the (right) developing pelvic diaphragm from the inside of the pelvis. In the lower right corner is the os sacrum. In the upper left corner, the symphysis pubica can be seen.

The left and right corpus ossis pubis meet in the midline and form the symphysis pubica (*2*). The rectum and sinus urogenitalis, together forming part of the corpus pelvinum, have been left out in this reconstruction to give a better view of the muscles of the pelvic diaphragm.

The reconstruction of the m. obturatorius internus is shown in red. The ventral to dorsal extension from corpus ossis pubis to the corpus ossis ischii is visible. The dorsal extension of the m. obturatorius internus is about as far as in the 34-mm fetus. Dorsally it does not completely cover the ramus and corpus ossis ischii. The muscular volume of the m. obturatorius internus itself also cannot be seen leaving the pelvis to insert on the head of the femur. However, the fascia of the muscle does but is not reconstructed. On the ventral and cranial side, the m. obturatorius does not reach the lower margin of the ramus superior ossis pubis. In the 34-mm embryo, the m. obturatorius internus fills in approximately three-quarters of the foramen obturatum, being absent at its cranial part. On the ventral side, the m. obturatorius internus covers about half the foramen obturatum (Fig. 39). A canalis obturatorius as such is not present. When comparing the extension of the m. obturatorius internus to the 34-mm fetus, it is clear that the development in

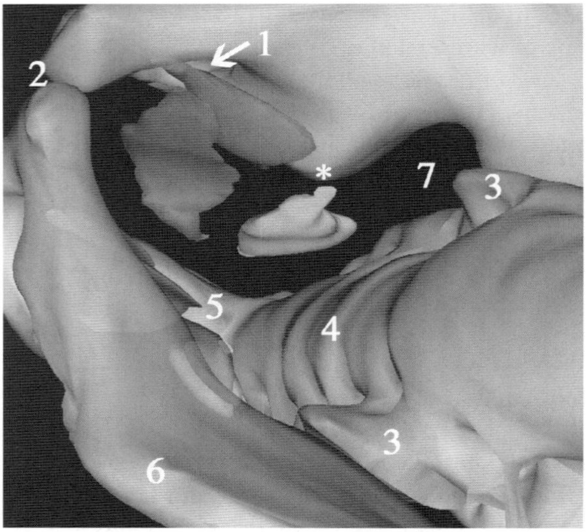

Fig. 38 View from above on the pelvic diaphragm. *1*, foramen obturatum; *2*, symphysis pubica; *3*, pars lateralis os sacrum; *4*, os sacrum; *5*, os coccygis; *6*, ala ossis ilii; *7*, foramen ischiadicum majus; *, future spina ischiadica; *red*, m. obturatorius internus; *blue*, m. levator ani; *green*, m. coccygeus (full color version of this illustration at the end of the book)

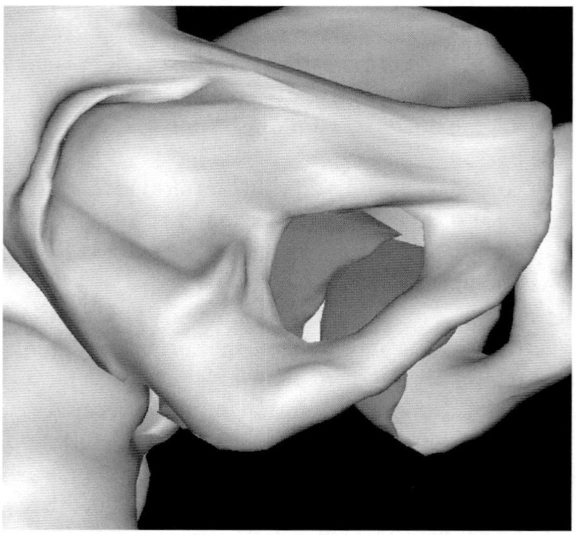

Fig. 39 View of the right foramen obturatum. The m. obturatorius internus is reconstructed in *red*. Note that the muscle does not cover the frontal half and cranial quarter of the foramen obturatum (full color version of this illustration at the end of the book)

the dorsal direction is roughly the same. The ventral and cranial direction is less developed compared to the 34-mm fetus.

The reconstruction of the m. levator ani is shown in dark blue. Near the symphysis pubica it covers the m. obturatorius internus for about a quarter of its length, comparable to the 34-mm fetus. More distally, the m. levator ani covers the m. obturatorius internus for about half its length. The muscular volume of the m. levator ani in this 28-mm embryo does not reach the level of the future spina ischiadica. The border between the levator ani and the obturatorius internus near its origin at the level of the symphysis pubica is in fact the ATLA, running from the inner aspect of the ramus inferior ossis pubis to the future developing spina ischiadica. The m. levator ani has not yet developed far enough dorsally to reach most of the ATLA. Whether the ATLA already reaches the region of the future spina ischiadica cannot be distinguished in the available sections. This reconstruction of the m. levator ani also makes it distinctly clear that the levator ani consists of a single muscle that covers the frontal half of the pelvic outlet. The dorsal extension of the muscle is less developed compared to the 34-mm fetus, especially at the level of the m. coccygeus. In contrast to the 43-mm and 50-mm fetuses, it does not yet have the typical triangular shape. It is in this sense comparable to the 34-mm fetus.

The m. coccygeus is reconstructed in green. Its origin is at the level of the future spina ischiadica and its direction is towards the surface of the last sacral and first coccygeal vertebrae. The m. coccygeus covers the dorsal half of the fetal pelvic outlet. When comparing the m. coccygeus to the 34-mm fetus, it is clear that the muscle volume shows slightly less development. The space between the m. coccygeus and the surface of the os sacrum is a bit broader. This is due to the fact that the dorsal extension of the m. coccygeus in this 28-mm embryo is less than in the 34-mm fetus. The m. levator ani and the m. coccygeus are two separate muscles, distinguishable from each other by a space filled with connective tissue.

The m. sacrococcygeus ventralis is not reconstructed in this 28-mm embryo. It does not cover any part of the actual pelvic outlet and is therefore not considered to belong to the pelvic diaphragm (see the reconstruction of the 50-mm fetus for an example).

3.1.7 Three-Dimensional Reconstruction Frontal and Caudal Views of the Pelvic Diaphragm

Figure 40 shows a complete frontal picture of the reconstruction. It is clear that the pelvic outlet of this 28-mm embryo is also projected in the frontal plain, thus pointing forwards. The sharp angle of the os coccygis to the os sacrum in this embryo is clearly visible in Fig. 40. The forward projection of the pelvic diaphragm is identical to the older fetuses.

The m. obturatorius internus is reconstructed in red. It covers the lower three-quarters and dorsal half of the foramen obturatum (see also Fig. 39). In Fig. 40, it is also visible that the m. obturatorius internus does only cover the lower part of

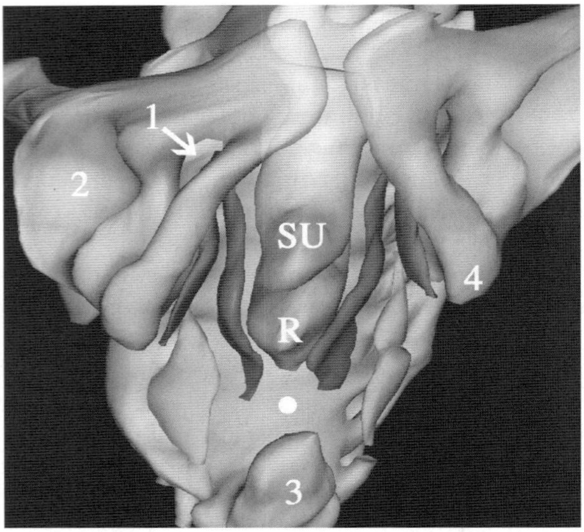

Fig. 40 Frontal view of the pelvic diaphragm. *R*, rectum; *SU*, sinus urogenitalis; *1*, foramen obturatum; *2*, acetabulum; *3*, os coccygis; *4*, ramus ossis ischii; *white dot*, future ligamentum anococcygeum; *red*, m. obturatorius internus; *blue*, m. levator ani; *green*, m. coccygeus. Note the absence of the curving between the sinus urogenitalis and the rectum (the future centrum tendineum) (full color version of this illustration at the end of the book)

the foramen obturatum. A clearly defined canalis obturatorius is not present, as in the later stages.

The m. levator ani is shown in dark blue. It runs from the corpus ossis pubis and lower margin of the ramus inferior ossis pubis to the os coccygis. The left and right m. levator ani surround the corpus pelvinum, which is reconstructed in yellow.

The left and right levator ani are each separate muscles and are curved inwards and towards each other at a level between the rectum and the os coccygis. This curving, however, is more pronounced than in the 34-mm fetus. Behind the rectum, in front of the os coccygis, the left and right levator ani will eventually reach the developing ligamentum anococcygeum. The distance between the rectum and the tip of the os coccygis is less than it is in the 34-mm fetus. This is because the os coccygis is slightly more curved towards the rectum. The distance between the left and right m. levator ani at this point is the same as in the 34-mm fetus. More interesting is the absence of inward curving of the m. levator ani between the sinus urogenitalis and the rectum. Apparently, the formation of a future centrum tendineum has not started.

Closure of the pelvic diaphragm in the midline is not present. Compared to the 34-mm fetus, closure of the pelvic diaphragm in the midline is equally advanced in this 28-mm embryo. The insertion of the m. levator ani onto the surface of the os coccygis is also not yet completed. A small gap is still present, formed by connective tissue that is not reconstructed.

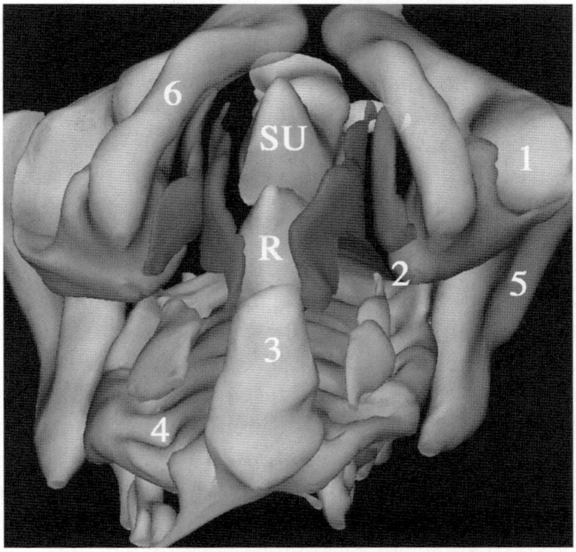

Fig. 41 View of the pelvic diaphragm from below. *R*, rectum; *SU*, sinus urogenitalis; *1*, acetabulum; *2*, ramus ossis ischii at level of developing spina ischiadica; *3*, os coccygis; *4*, os sacrum; *5*, os ilium (ala ossis ilii); *6*, ramus inferior ossis pubis; *red*, m. obturatorius internus; *blue*, m. levator ani; *green*, m. coccygeus (full color version of this illustration at the end of the book)

The m. coccygeus, reconstructed in green, appears to be slightly thicker than the m. levator ani. The volume of the m. coccygeus is about the same as in the 34-mm fetus, only slightly smaller. The position of the m. coccygeus is more lateral in relation to the m. levator ani. The two muscles show what looks like a roof tile formation. This is the same spatial position as in the other described fetal stages.

A close look at Fig. 41 suggests that the pelvic diaphragm consists of four quadrants, as it does in the older fetuses. The ventral left and ventral right quadrants are filled by the left and right m. levator ani and the dorsal quadrants by the m. coccygeus. The m. coccygeus again appears slightly smaller than in the 34-mm fetus. It does not fill the dorsal quadrants as much as it does in the later stages.

3.2 Stage 21: 22 mm (WK 7832) and 23 mm (WK 1256, WK 548, 5249), ± 53–54 days

3.2.1 General Features

These embryos with a CRL of 22 mm and 23 mm, respectively, are about 7.5 weeks (53–54 days) postconception. The embryos are in developmental stage 21 according to the subdivision by Streeter (1942–1951, described in Gribnau and Geijsberts 1981). At this stage, the separation of the fingers and toes starts. Distinctive external characteristics are: the distal phalanges of the fingers are already separated, the arms and legs are in a flexed position at the elbows and knees, the hands and feet of both sides are approaching each other and the tips of the toes are separated (Gribnau and Geijsberts 1981). The embryos are cut transverse in 10-μm sections and stained with HE (see Table 2). The description of stage 21 is given by the 23-mm embryo (WK 1256). The other embryos are in essence the same. If necessary, the other stage-21 embryos will be separately discussed.

Figures 42 and 43 shows a section through the 23-mm embryo. The developing kidneys (metanephros) are clearly visible. In between the kidneys is the aorta. Lateral and caudal to the kidneys, the urogenital ridge can be seen with the mesonephros and the developing testis. The embryo is male.

The direction of the sections is oblique–horizontal. Due to the strong curving of the embryo at this stage, the sections through the pelvis also show in part the upper urogenital system. The orientation of the sections through the pelvis is more frontal. Fig. 42 shows clearly that the sinus urogenitalis is embedded in dense connective tissue and mesenchyme. Far more dense than in the older embryos and fetuses. Fat is not present.

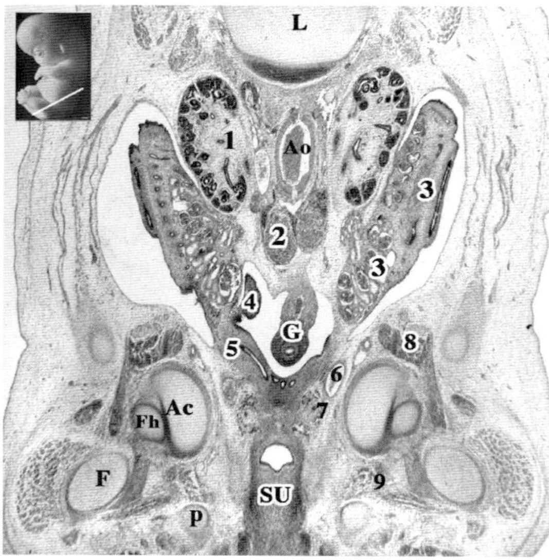

Fig. 42 *Inset*: level of section indicated by a *white line*. *Ac*, acetabulum; *Ao*, aorta; *F*, femur; *Fh*, caput femoris; *G*, gut; *L*, lumbar vertebra; *p*, os pubis (ramus inferior); *SU*, sinus urogenitalis; *1*, metanephros (kidney); *2*, chromaffin tissue; *3*, mesonephros; *4*, testis; *5*, mesonephric duct (Wolffian duct); *6*, arteria umbilicalis; *7*, ureter; *8*, m. psoas major; *9*, m. obturatorius externus

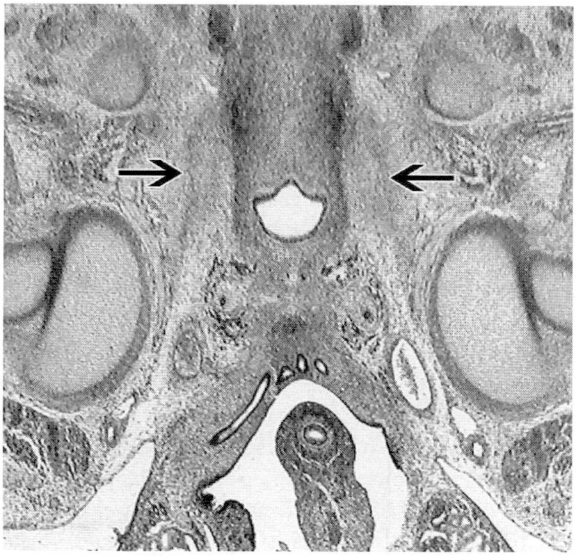

Fig. 43 Enlarged detail of Fig. 42, but upside down. The *black arrows* point towards the arcus tendineus levatoris ani, running from in between the sinus urogenitalis and ramus inferior ossis pubis towards the pelvic aspect of the lower end of the acetabulum. The ATLA can be seen as a *dark gray band*

The development of the bony pelvis is less advanced compared to the older embryos. All the individual bones that constitute the pelvis are outlined in cartilage. The corpus ossis ilii is present, but the ala ossis ilii has just started to develop. The reconstruction of the pelvis, shown in Fig. 46, reveals small developing ala ossis ilii on the left and right sides. The corpus ossis pubis on both sides do not reach the midline. The symphysis pubica has not yet been formed. The ramus inferior and superior of the os pubis together with the os ischii surround the foramen obturatum. The tuber ischiadicum and the spina ischiadica are not present.

The os sacrum of the embryos is fully developed in cartilage. Fusion of the vertebrae is not yet present. The connection of the os sacrum to the os coxae is mesenchymal, attaching the pars lateralis of the os sacrum to the os coxae. Ligg. sacroiliaca interossea present in the older fetuses cannot be identified as such. The os coccygis is formed in cartilage and shows fusion of the third to fifth vertebra.

After evaluating the reconstruction of the cartilaginous pelvis of the 23-mm embryo, it is apparent, comparable to the later stages, that the typical angle at the level of the promontorium is not visible. There is even slight inward curving at the level of the promontorium. The os coccygis makes a sharp angle ventrally. Absence of the angle at the level of the promontorium and the sharp angle of the os coccygis seems responsible for the ventral position of the pelvic outlet present in this embryonic stage.

3.2.2 Musculus Obturatorius Internus

The difference in the m. obturatorius developmental stage is striking, compared to the later embryonic and fetal stages.

Running through the sections in a caudal direction, the nervus obturatorius appears. It can be seen leaving the pelvic cavity through the foramen obturatum. The point where it pierces the foramen obturatum is well below the lower margin of the ramus inferior ossis pubis, between the upper and middle third of the foramen. Similar to the later stages, it follows the ventral edge of the os ischii, but at a lower level. It runs towards the m. obturatorius externus.

At the lateral aspect of the foramen obturatum, the m. obturatorius externus is present. It covers the complete inner margins of the foramen and can be followed up to its insertion on the fossa trochanterica femoris. At the pelvic aspect of the foramen, there is no appearance of any muscle. The pelvic aspect is covered with connective tissue or mesenchyme. There is no sign of the presence of a membrana obturatoria. At the lower edge of the foramen, at the level of the inner aspect of the ramus ossis ischii, the m. obturatorius internus appears. It covers the pelvic aspect of the ramus ossis ischii. The muscle can be traced, leaving the pelvis to insert onto the trochanter major femoris. In this embryo there is muscle tissue all the way up to the insertion. This is in contrast to the later embryos and early fetal stages already described. It is not earlier than the 50-mm fetus that again the m. obturatorius internus shows muscle fibers all the way up to its insertion on the femur.

Between the medial aspects of both the corpus ossis pubis and the ventral and lateral sides of the allantois, a large area of dense tissue is present. In this tissue, mesenchymal formations are present that appear to bridge the allantois with the corpus ossis pubis. This mesenchyme is the developing endopelvic fascia and pubovesical ligaments.

At the beginning of the lower two-thirds of the foramen obturatum, a dense band of tissue arises at the level of the nervus obturatorius. It starts at the level of the acetabulum at the inner aspect of the corpus ossis ischii and stretches out towards the medial aspect of the ramus inferior ossis pubis. It does not reach the surface of the ramus inferior ossis pubis. It ends at the level where the sinus urogenitalis leaves the pelvic cavity. At that point it lies in between the ramus inferior ossis pubis and the sinus urogenitalis, separated by dense tissue. It is connected with the developing endopelvic fascia. This band of tissue is mesenchyme of the developing arcus tendineus levatoris ani (ATLA). Following the ATLA through the sections, muscles fibers start to appear at its ventral aspect, in between the ramus inferior ossis pubis and the sinus urogenitalis. This is the m. levator ani. At the moment that the m. levator ani appears, the m. obturatorius internus arises at the level of the ramus ossis ischii. This means that at this stage, the origin of the m. levator ani is not on the inner surface of the m. obturatorius internus, which was the case in later stages.

3.2.3
Musculus Levator Ani

As the m. levator ani originates, it is immediately clear that the m. levator ani is a single muscle.

Comparable to the later stages, the m. levator ani shows a gradual thickening at its ventral side. The ventral margin is surrounded by dense tissue that exists of connective tissue and mesenchyme. As already mentioned earlier, this embedding tissue can be followed towards the ventral and lateral sides of the sinus urogenitalis and farther cranial to the ventrolateral sides of the allantois. In front of the allantois, the tissue makes contact with the medial side of the ramus inferior ossis pubis and the allantois itself. So there is a connection, via mesenchyme, of the m. levator ani, the allantois, the sinus urogenitalis and the arcus tendineus levatoris ani. This tissue has to be the future endopelvic fascia, making contact with the ATLA, including the pubovesical ligaments. The urogenital diaphragm will develop in this area.

At its medial side, the m. levator ani is separated from the sinus urogenitalis by a broad band of loose connective tissue. There is no direct connection between the two structures. Following this border through the sections, it becomes smaller behind the sinus urogenitalis and in front of the hindgut. This separation of the m. levator ani by loose connective tissue is maintained all the way up to the disappearance of the m. levator ani in front of the tip of the os coccygis. So there is no direct connection between the sinus urogenitalis and hindgut and the m. levator ani. Behind the sinus urogenitalis, at the level of the hindgut, the ventral margin of the m. levator ani makes contact with the deep part of the external sphincter. At this stage, a separate deep and larger superficial part of the m. sphincter ani externus can be identified.

At its lateral side, the m. levator ani is first covered by connective tissue and mesenchyme that can be followed all the way up to the foramen obturatum. Following the muscle trough the sections, the m. obturatorius internus starts to appear at the dorsolateral side of the m. levator ani. They keep a close relationship near the upper margin of the ramus ossis ischii, but show no direct connection. Passing alongside the ramus ossis ischii the border between the m. levator ani and the m. obturatorius internus increases as the m. obturatorius internus pulls away, leaving the pelvis. The lateral side of the m. levator ani is now covered by the developing structures in the fossa ischiorectalis. Passing the hindgut, the m. levator ani is covered on its lateral side by connective tissue and fat of the developing buttocks.

Cranially, the dorsal side of the m. levator ani continues into the arcus tendineus levatoris ani. Following the dorsal margin through the embryo, the m. obturatorius internus appears. After the disappearance of the m. obturatorius internus, the dorsal margin is constituted of loose connective tissue until the m. levator ani disappears in front of the os coccygis. Near the point where the m. obturatorius leaves the pelvis, a third muscle appears at the dorsal margin of the corpus ossis ischii. This muscle is the m. coccygeus.

At its most caudal margin, the m. levator ani is lost in dense tissue that can be followed up to the tip of the os coccygis. This tissue is the developing ligamentum anococcygeum as it forms a connection between the dorsal end of the external sphincter (deep part), the m. levator ani (of both sides) and the os coccygis. Figure 45 gives a good detail of the levator ani and its relationship to the ligamentum anococcygeum.

3.2.4 Musculus Coccygeus

At the surface of the most dorsal part of the corpus ossis ischii, the m. coccygeus originates. No formation of a future spina ischiadica is present at this stage. The origin of the m. coccygeus is at the level where the m. obturatorius internus leaves the pelvis. In adults, this point is just below the spina ischiadica. A distinct presence of the ATLA is not noted in the 22-mm embryo. In the 23-mm WK 548 embryo, the developing ATLA is identified (Fig. 44).

After it originates, the m. coccygeus rapidly increases in volume as it progresses towards the os coccygis. At this embryonic stage, the m. coccygeus already appears to constitute of a dorsal and ventral part. In contrast to the later embryos and fetuses, this subdivision is hardly visible, which can be noted by the different orientation by which the muscle fibers are cut. The dorsal part is the future ligamentum sacrospinale and is bigger than the ventral part, being the actual m. coccygeus. At some point in the m. coccygeus of the 23-mm WK548 embryo, the division of the m. coccygeus is more easily discerned (Fig. 45).

Following the dorsal part of the m. coccygeus through the embryo, it is lost in connective tissue that is fixed to the lateral surface of the os sacrum.

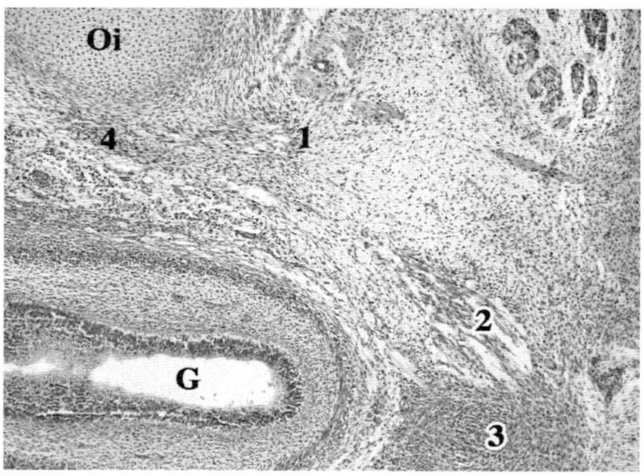

Fig. 44 Detail of 23-mm WK548 embryo. Left side spinal column. *G*, hindgut; *Oi*, ramus/corpus ossis ischii; *1*, m. coccygeus; *2*, m. levator ani; *3*, deep part of m. sphincter ani externus; *4*, part of arcus tendineus levatoris ani. Note the fixation of the m. levator ani to the external anal sphincter

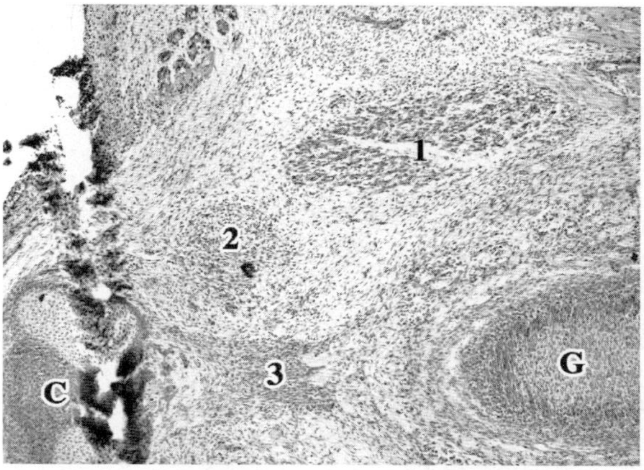

Fig. 45 Detail of 23-mm WK548 embryo. *Black artifact* is cutting damage. Left side is ventral. *C*, os coccygis; *G*, hindgut; *1*, m. coccygeus; *2*, m. levator ani; *3*, ligamentum anococcygeum. The m. coccygeus is divided into ventral and dorsal parts. The levator ani comes in to close contact with the ligamentum anococcygeum on its way to insert on the lateral side of the os coccygis

The ventral part travels towards the midline and can be followed nearly all the way up to the insertion on the os coccygis. Connection of the m. coccygeus to the os coccygis is made by connective tissue that is far less definable as fascia than it is in the 34-mm fetus (Fig. 28).

3.2.5 Musculus Sacrococcygeus Ventralis

The m. sacrococcygeus ventralis cannot be identified in this embryo.

3.2.6 Three-Dimensional Reconstruction Oblique View of the Pelvis

This overview of the reconstructed pelvis (Fig. 46) gives a good impression of the degree of development of the pelvis. It is beyond the scope of this thesis to describe the development of the bony pelvis itself. Yet it is interesting to compare the pelvis of this 23-mm embryo to the pelvis of the 50-mm fetus (Fig. 11). Note for instance the relatively small volume of the pelvis compared to the spinal column. The ala ossis ilii have started to develop, extending as a small rim to the front. The left and right corpus ossis pubis do not meet in the midline. There is no symphysis pubica. The foramen obturatum is completely formed, but is relatively small. The acetabulum is outlined, but it is not as deep as in the 50-mm fetus. Especially note the relatively large volume of the os coccygis.

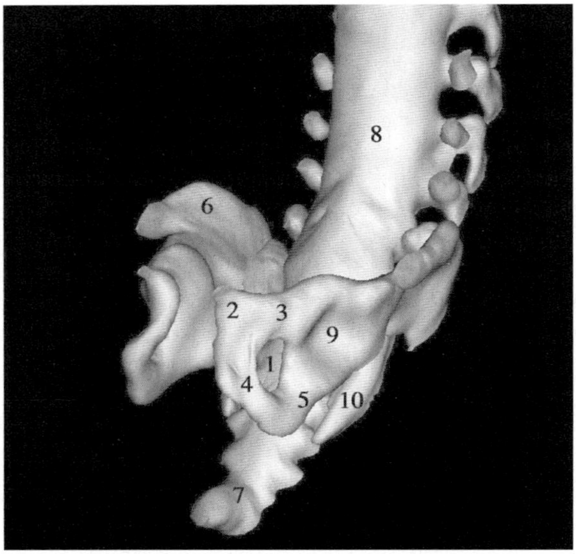

Fig. 46 Oblique view of the reconstructed pelvis. *1*, foramen obturatum; *2*, corpus ossis pubis; *3*, ramus superior ossis pubis; *4*, ramus inferior ossis pubis; *5*, ramus ossis ischii/corpus ossis ischii; *6*, developing ala ossis ilii; *7*, os coccygis; *8*, spinal column; *9*, acetabulum; *10*, lateral margin of os sacrum

3.2.7
Three-Dimensional Reconstruction Oblique View from Above on the Pelvic Diaphragm

This picture (Fig. 47) of the three-dimensional reconstruction is taken from above, showing the (right) developing pelvic diaphragm from the inside of the pelvis. In the lower right corner is the os sacrum, in the upper left corner the os pubis. The pelvic rim is open in the front. The reconstructed sinus urogenitalis and hindgut have been left out for a better view.

The reconstruction of the m. obturatorius internus is shown in red. Figure 48 shows a more detailed overview of the foramen obturatum and the position of the m. obturatorius internus. It now becomes clear that the m. obturatorius internus does not reach the foramen obturatum in this stage of its embryological development. It is situated on the inner side of the ramus and corpus ossis ischii (Fig. 47). It can be seen leaving the pelvis at the dorsal edge of the ramus ossis ischii and at the caudal part of the corpus ossis ischii. In the older embryos and fetuses, the m. obturatorius internus extends farther in the direction of the corpus ossis pubis. The dorsal extension and the insertion on the trochanter major femoris is identical to the later stages.

The reconstruction of the m. levator ani is shown in dark blue. It lies in front and does not cover the m. obturatorius internus anywhere. The ventral to dorsal extension does not change in the direction of the os coccygis, except for the

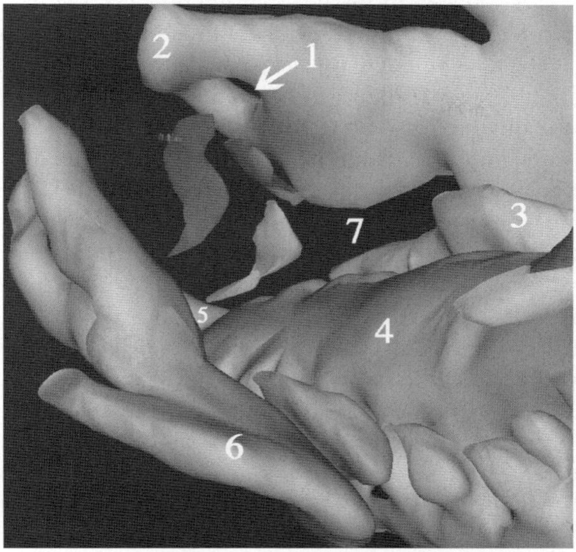

Fig. 47 View from above on the pelvic diaphragm. *1*, foramen obturatum; *2*, corpus ossis pubis; *3*, pars lateralis os sacrum; *4*, os sacrum; *5*, os coccygis; *6*, ala ossis ilii; *red*, m. obturatorius internus; *blue*, m. levator ani; *green*, m. coccygeus (full color version of this illustration at the end of the book)

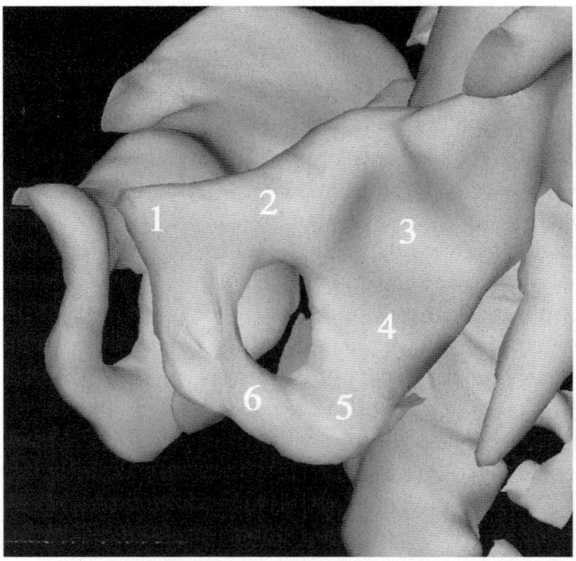

Fig. 48 Close view of the left foramen obturatum. *1*, corpus ossis pubis; *2*, ramus superior ossis pubis; *3*, acetabulum; *4*, corpus ossis ischii; *5*, ramus ossis ischii; *6*, ramus inferior ossis pubis; *red*, m. obturatorius internus (full color version of this illustration at the end of the book)

last part where it disappears in the ligamentum anococcygeum. The m. levator ani shows the same curving around the sinus urogenitalis as it does in the later stages, caused by the attachment to the deep part of the m. sphincter ani externus. The sphincter is not reconstructed. The position of the m. levator ani relative to the os pubis is different compared to the later stages. The ventral margin of the levator ani is farther away from the os pubis. The gap is filled by connective tissue and mesenchyme forming the endopelvic fascia and part of the diaphragma urogenitalis. The most cranial margin of the m. levator ani is smaller. This cranial margin is in fact the ATLA, which is not followed as much dorsally as it does in the older embryos and fetuses. The levator ani closes a smaller part of the ventral quadrant of the pelvic outlet compared to the 28-mm embryo.

The m. coccygeus is reconstructed in green. Its origin is at the level of the future spina ischiadica and its insertion is on the surface of the last sacral and first coccygeal vertebrae. It has roughly the same shape compared to the 28-mm embryo and 34-mm fetus. The reconstruction shows how the cranial part of the m. coccygeus extends dorsally towards the os sacrum. This is the part of the m. coccygeus that will become the ligamentum sacrospinale. The ventral to dorsal extension of the m. coccygeus becomes much smaller towards the os coccygis. It actually shows that the development of the m. coccygeus is less advanced, especially in the ventral direction towards the m. levator ani. The border between the m. levator ani and m. coccygeus is broader. The m. coccygeus closes a smaller part of the dorsal quadrant of the pelvic outlet than it does in the 28-mm embryo.

3.2.8 Three-Dimensional Reconstruction Frontal and Caudal View of the Pelvic Diaphragm

Figure 49 shows a complete frontal picture of the reconstruction. As in the older embryos and fetuses, the pelvic outlet is projected in a frontal plain. The sharp angle of the os coccygis to the os sacrum is less striking than it is in the 28-mm embryo, but comparable to the 34-mm fetus.

The m. obturatorius internus is reconstructed in red. Note the position of the muscle on the pelvic surface of the ramus and corpus ossis ischii (Fig. 50), extending less cranially compared to the 28-mm fetus.

The reconstruction of the m. levator ani is shown in dark blue. It runs from the lower margin of the upper part of the ramus inferior ossis pubis towards the os coccygis. The upper margin of the levator ani is therefore lower than in the 28-mm embryo, where it extends towards the lower part of the corpus ossis pubis. Also note that the m. levator ani extends farther cranial than the m. obturatorius internus (Fig. 50), whereas in the 28-mm embryo the m. obturatorius internus extends farther. The left and right m. levator ani surround the corpus pelvinum, which is reconstructed in yellow. Both the left and the right m. levator ani are separate muscles and have a single curve around the corpus pelvinum. In the 28-mm embryo, there was a second curving behind the rectum (hindgut) and in later stages also between the sinus urogenitalis and the rectum. Between the

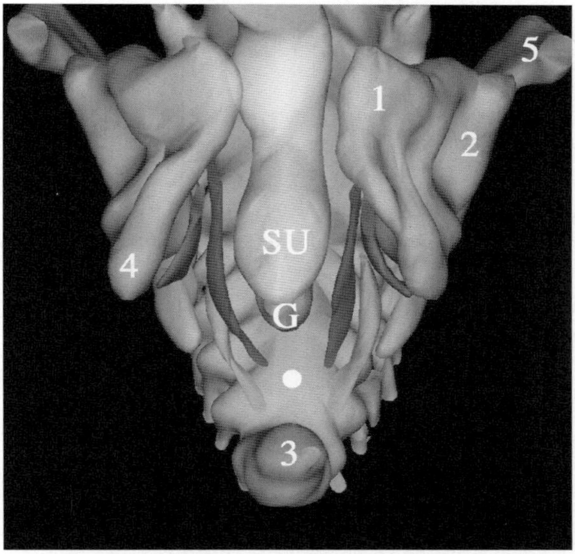

Fig. 49 Frontal view of the pelvic diaphragm. *G*, hindgut; *SU*, sinus urogenitalis; *1*, corpus ossis pubis; *2*, acetabulum; *3*, os coccygis; *4*, ramus ossis ischii; *5*, ala ossis ilii; *white dot*, future ligamentum anococcygeum; *red*, m. obturatorius internus; *blue*, m. levator ani; *green*, m. coccygeus (full color version of this illustration at the end of the book)

hindgut and the os coccygis, the left and right levator ani curve towards the midline, reaching the developing ligamentum anococcygeum. The distance between the hindgut and the os coccygis is bigger than in the 28-mm, but comparable to the 34-mm fetus. Closure of the pelvic diaphragm in the midline is not present. Compared to the 28-mm embryo, closure of the pelvic diaphragm in the midline is less advanced in this 23-mm embryo. The insertion of the m. levator ani onto the surface of the os coccygis is also not yet completed. A small gap, but bigger than in the 28-mm embryo, is still present and is formed by connective tissue that is not reconstructed.

The m. coccygeus is reconstructed in green. It appears to be as thick as the m. levator ani. Its total volume is less than in the 28-mm embryo. In contrast to the older stages, the position of the m. coccygeus is in the same direction as the m. levator ani. In the older stages, the m. coccygeus is situated more oblique with its dorsal part more lateral. This is due to the fact that the os sacrum in this 23-mm embryo is smaller. Growing in size makes the dorsal part of the m. coccygeus extend more laterally. The m. coccygeus in total is situated lateral from the m. levator ani, which is comparable to the later stages.

Figure 50 suggests that the pelvic diaphragm consists of four quadrants, as it does in the older embryos and fetuses. The ventral left and ventral right quadrants are filled by the left and right m. levator ani and the dorsal quadrants by the m. coccygeus. The m. coccygeus again appears smaller than in the 28-mm embryo. It does not fill the dorsal quadrants as much as it does in the later stages.

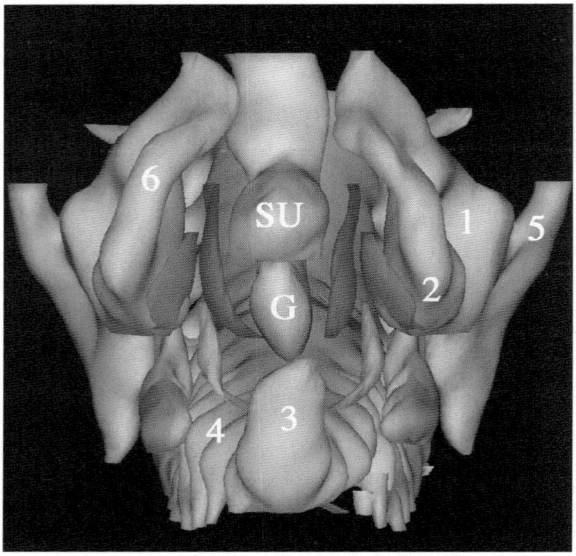

Fig. 50 View of the pelvic diaphragm from below. *G*, hindgut; *SU*, sinus urogenitalis; *1*, acetabulum; *2*, ramus ossis ischii at level of developing spina ischiadica; *3*, os coccygis; *4*, os sacrum; *5*, os ilium (ala ossis ilii); *6*, ramus inferior ossis pubis; *red*, m. obturatorius internus; *blue*, m. levator ani; *green*, m. coccygeus (full color version of this illustration at the end of the book)

3.3 Stage 20: 20 mm (WK7996, WK7262, WK7993, WK1423) and 21 mm (WK2287, WK8071), ± 52 days

3.3.1 General Features

These embryos with a CRL of 20 mm and 21 mm, respectively, are approximately 7.5 weeks (53–54 days) postconception. The embryos are in developmental stage 20 according to the subdivision by Streeter. The embryos are cut transverse in 10-μm sections and stained with HE (see Table 2). To make the description of stage 20 readable, only the 20-mm WK7262 embryo is described. The other embryos are in essence the same. If necessary, the other stage 20 embryos will be discussed separately.

Figure 51 shows a section through the 20-mm WK7993 embryo. Although this particular embryo is slightly macerated, it still gives a good overview of the developing pelvis. Studying the sections through the mesonephros, the embryo appears to be male. The section shown in Fig. 51 is a section well below the level of the mesonephros and metanephros. The section is slightly oblique. Since the pelvic outlet in this stage of development is pointed in a forward direction (comparable to all the later stages), the dorsal end of the section is cut through the lower lumbar spinal column and the ventral end through the level of the acetabulum.

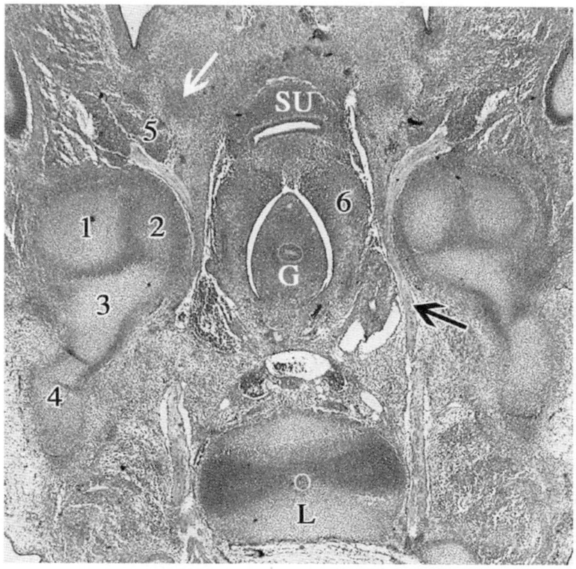

Fig. 51 *G*, hindgut; *L*, lumbar vertebra; *SU*, sinus urogenitalis; *white arrow*, ramus inferior ossis pubis; *black arrow*, nervus obturatorius; *1*, caput femoris; *2*, corpus ossis ischii; *3*, corpus ossis ilii; *4*, ala ossis ilii; *5*, m. obturatorius externus; *6*, embryonic mesothelium

The mesenchyme is more concentrated than in the 23-mm embryo. It is more compact and shows a lower level of development. There is hardly any connective tissue present, for example. From behind the hindgut and radiating around it, the mesothelium of the sinus urogenitalis is visible, containing the paramesonephric and mesonephric ducts as well as small blood vessels. Also note the relatively large volume of the nervus obturatorius, which can be followed from the vertebral column to the m. obturatorius externus. It can be seen leaving the pelvis, curving around the corpus ossis ischii.

The pelvis shows less development than the 23 mm embryo. The separately developing corpus ossis pubis, corpus ossis ischii and corpus ossis ilii are visible. The ala ossis ilii shows a similar development as in the stage-21 embryos. Compared to the 23-mm embryo, especially the ramus inferior ossis pubis is less developed. It does not show a cartilaginous pattern yet, but is outlined in clustered mesenchymal cells. The foramen obturatum is therefore just present, not completely surrounded by cartilage. The ramus superior ossis pubis is outlined in cartilage. Like the 23-mm embryo, the left and right os pubis have not reached the midline.

The development of the os sacrum is comparable to stage 21. It is completely outlined in cartilage. The connection of the pars lateralis of the os sacrum to the os ilium is made by mesenchyme. The os coccygis is present, but is not distinguishable from the os sacrum, other than being the last vertebrae of the spinal column. The presence of fusion of the last vertebrae cannot be concluded from the available material.

No reconstruction has been made of the pelvis and pelvic diaphragm. The development of the pelvic diaphragm will be discussed and illustrated with photographs. A reconstruction of the developmental stage of the separate muscles appears to have no extra value over the reconstructions of the 23-mm embryo. This will become clear after the description of these 20-mm embryos.

3.3.2 Musculus Obturatorius Internus

Running through the sections in a caudal direction, the nervus obturatorius appears. It can be seen leaving the pelvic cavity through the foramen obturatum. The point where it pierces the foramen obturatum is well below the lower margin of the ramus inferior ossis pubis, somewhere between the upper and middle third of the foramen. Similar to the later stages, it follows the ventral edge of the os ischii, but at a lower level. It runs towards the m. obturatorius externus. Since the development of the os pubis is less advanced, it can now be concluded that the location of the nervus obturatorius at a lower level is caused by a smaller ramus superior ossis pubis. The nervus obturatorius seems to have a fixed position and the ramus superior ossis pubis will displace itself towards the nerve. Figure 52 shows a photograph of the 20-mm WK7262 embryo at the upper level of the foramen obturatum. In this embryo, the ramus inferior ossis pubis shows a cartilaginous pattern. The development of cartilage in the ramus inferior ossis pubis thus shows some variation in stage-20 embryos. Presumably the embryonic stage 20 is the stage where chondrification of the ramus inferior begins.

At the lateral aspect of the foramen obturatum, the m. obturatorius externus is present. It covers the complete inner margins of the foramen and can be followed

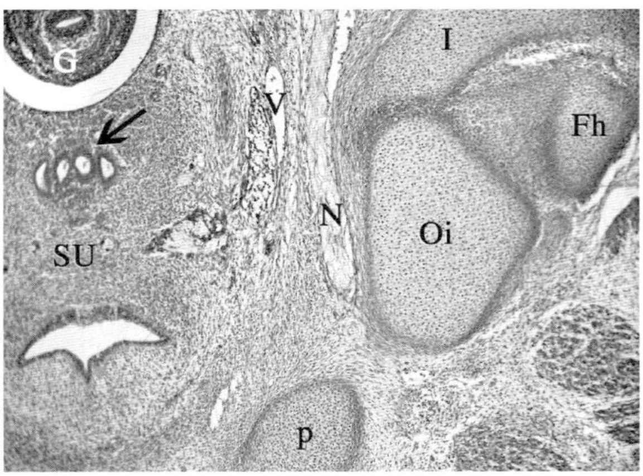

Fig. 52 *Fh*, caput femoris; *G*, hindgut; *I*, corpus ossis ilii; *N*, nervus obturatorius; *Oi*, corpus ossis ischii; *p*, ramus inferior ossis pubis; *SU*, sinus urogenitalis; *V*, a/v obturatoria; *black arrow*, mesonephric and paramesonephric ducts

up to its insertion on the fossa trochanterica femoris. The pelvic aspect of the foramen obturatum is covered with mesenchyme. There is no sign of presence of a membrana obturatoria. At the lower edge of the foramen, at the level of the inner aspect of the ramus ossis ischii, the m. obturatorius internus appears (Fig. 53). It covers the pelvic aspect of the ramus ossis ischii, comparable to the 23-mm embryos. The muscle can be traced leaving the pelvis to insert on the trochanter major femoris (Fig. 54). The m. obturatorius internus shows muscle fibers all the way up to its insertion onto the trochanter major femoris.

In contrast to embryonic stage 21, no clear arcus tendineus levatoris ani can be identified. In Fig. 53, some sign of development of the levator ani fascia is present, as a dense band of mesenchyme that can be followed from the m. levator ani, over the m. obturatorius internus, towards the dorsal margin of the corpus ossis ischii. This figure also shows that the m. obturatorius lies dorsolateral from the m. levator ani. Similar to the 23-mm embryos, the m. levator ani does not cover the m. obturatorius internus.

In these 20-mm and 21-mm embryos, no sign of a developing endopelvic fascia can be found, nor is there any sign of developing pubovesical ligaments.

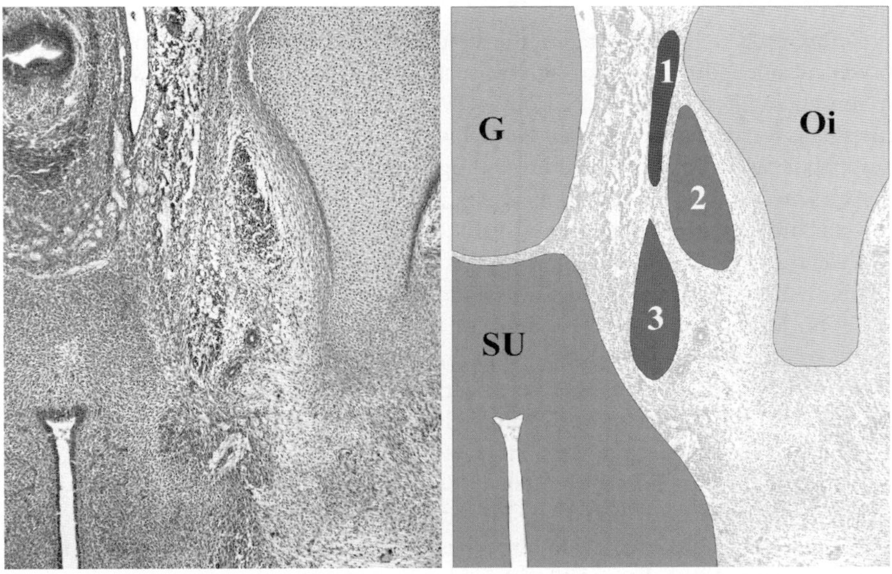

Fig. 53 *G*, hindgut; *Oi*, corpus ossis ischii; *SU*, sinus urogenitalis; *1*, developing fascia of the m. levator ani; *2*, m. obturatorius internus; *3*, m. levator ani. A dense band of mesenchyme can be followed from the m. levator ani, over the m. obturatorius internus, towards the dorsal margin of the corpus ossis ischii. This is the developing fascia of the m. levator ani. It runs towards the future ATLA at the mediodorsal surface of the corpus ossis ischii (at level of the *1*). A little below this level, the m. coccygeus will appear

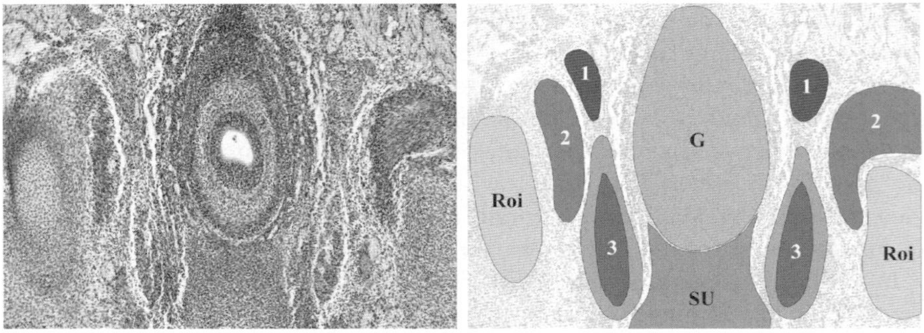

Fig. 54 *G*, hindgut; *Roi*, ramus ossis ischii; *SU*, sinus urogenitalis; *1*, m. coccygeus; *2*, m. obturatorius internus; *3*, m. levator ani. The m. obturatorius internus on the embryos right side curves around the dorsal margin of the corpus ossis ischii to insert on the trochanter major femoris. Note that a thick layer of mesenchyme, developing fascia, covers the m. levator ani

3.3.3 Musculus Levator Ani

As the muscle originates, it is immediately clear that the m. levator ani is a single muscle.

Comparable to the later stages, the m. levator ani shows a gradual thickening at its ventral side following it through the sections. The ventral margin is surrounded by mesenchyme, making close contact with the sinus urogenitalis (Fig. 53). In this stage, it is no longer possible to tell whether this mesenchyme is developing endopelvic fascia or part of the developing diaphragma urogenitalis.

At its medial side, the m. levator ani is separated from the sinus urogenitalis by a broad band of loose tissue containing blood vessels. There is no direct connection between the two structures. Following this border through the sections, it becomes smaller behind the sinus urogenitalis and in front of the hindgut. This separation of the m. levator ani and the content of the corpus pelvinum is maintained all the way up to the disappearance of the m. levator ani in front of the tip of the os coccygis. So there is no direct connection between the sinus urogenitalis, the hindgut and the m. levator ani. Behind the sinus urogenitalis, at the level of the hindgut, the ventral margin of the m. levator ani makes contact with the deep part of the external sphincter. At this stage, a separate deep and larger superficial part of the m. sphincter externus can be identified (Fig. 55).

At its lateral side, the m. levator ani is first covered by connective tissue and mesenchyme that can be followed all the way up to the foramen obturatum. Following the muscle through the sections, the m. obturatorius internus starts to appear at the dorsolateral side of the m. levator ani (Fig. 53). They keep a close relationship near the upper margin of the ramus ossis ischii, but show no direct connection. Passing alongside the ramus ossis ischii, the border between the m. levator ani and the m. obturatorius internus increases as the m. obturatorius internus pulls away, leaving the pelvis (Fig. 54). The lateral side of the m. levator ani is now formed by

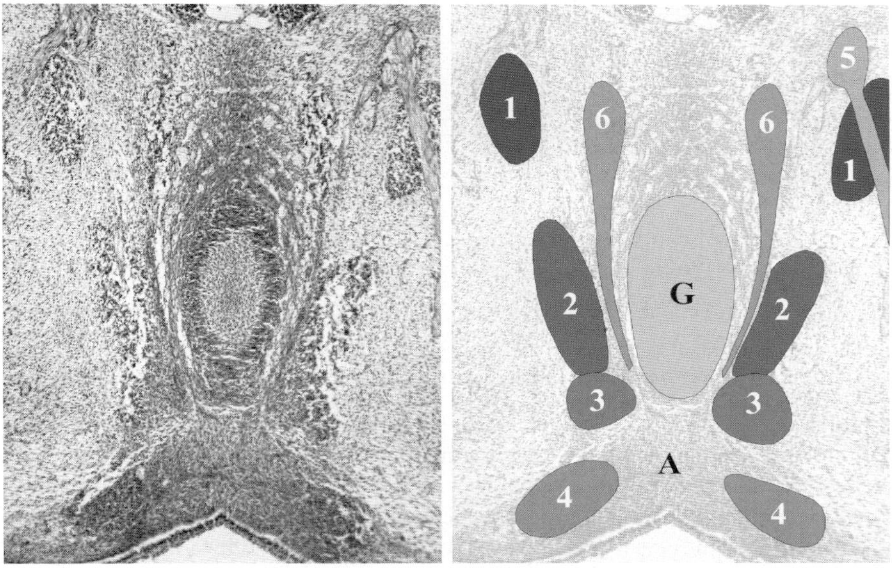

Fig. 55 *A*, anal membrane; *G*, hindgut; *1*, m. coccygeus; *2*, m. levator ani; *3*, deep part of m. sphincter ani externus; *4*, superficial part of m. sphincter ani externus; *5*, branch of nervus pudendus; *6*, longitudinal muscle layer rectum. The lower end of the picture is the ventral side of the embryo. The anal membrane is visible in between the superficial part of the external anal sphincter. The m. levator ani is attached to the deep part of the external anal sphincter. The m. coccygeus on the left side is divided into a ventral and dorsal part by a branch of the nervus pudendus

the developing fossa ischiorectalis. These structural relations are comparable to the embryonic stage 21.

Behind the hindgut, the m. levator ani travels towards the os coccygis. The connection with the m. sphincter ani externus is lost, as the sphincter disappears. The m. levator ani ends in the developing ligamentum anococcygeum. At this point, both the left and the right m. levator ani reach the midline, without making direct contact. The ligamentum anococcygeum is intermediate between the left and right m. levator ani. This ligament can be traced in these embryos, linking the deep part of the m. sphincter ani externus and the m. levator ani with the os coccygis (Fig. 56).

3.3.4 Musculus Coccygeus

At the surface of the most dorsal part of the corpus ossis ischii, just above the point where the m. obturatorius internus leaves the pelvis, the m. coccygeus appears. At its point of origin, no sign of a developing spina ischiadica can be noted, nor can the arcus tendineus levatoris ani be identified in this region.

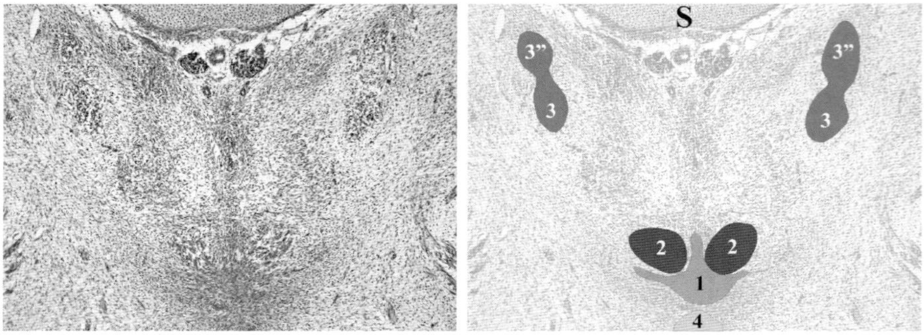

Fig. 56 *S*, os sacrum; *1*, ligamentum anococcygeum; *2*, m. levator ani; *3*, m. coccygeus; *3"*, dorsal part of m. coccygeus (ligamentum sacrospinale); *4*, anal seam

Following the m. coccygeus towards the os coccygis, its volume increases but to a lesser extent compared to the older embryos. A subdivision in a ventral and dorsal part is present, although it is difficult to delineate the separate parts. The subdivision becomes clear when the m. coccygeus is pierced by a nerve (Fig. 55). Below this level, the subdivision of the m. coccygeus is better outlined (Fig. 56). The dorsal part of the m. coccygeus travels towards the os sacrum and will form the ligamentum sacrospinale. The ventral part, or actual m. coccygeus, can be followed through the embryo in the direction of the os coccygis. It is lost in mesenchyme that reaches the surface of the os coccygis.

3.3.5 Musculus Sacrococcygeus Ventralis

The m. sacrococcygeus ventralis cannot be identified in this embryo.

3.4 Stage 18–19: 17 mm (WK359, WK4132, WK5032, WK7435), ± 49 days

3.4.1 General Features

These embryos with a CRL of 17 mm are roughly 7 weeks postconception. They are in developmental stage 18 according to Streeter. The embryos are cut transverse and stained with HE (see Table 2).

In this stage, we were not able to determine the sex type of the embryos by studying the mesonephros and paramesonephros. This is in accordance with embryological textbooks, which indicate that it is not possible to determine the sex type of embryos up to 17 mm CRL on the basis of the gonads (Hamilton et al. 1966). Therefore we looked in the embryos for the presence of the sex chromatin (Barr test), as described earlier. No sex chromatin was present in all the four embryos, so all the embryos are considered male. This could well be a coincidence or a false-negative result of the Barr test. Yet, from the embryos described earlier, it is already clear that the pelvic diaphragm develops independently of the sinus urogenitalis and therefore independently of its sex type.

The development of the pelvis is less advanced compared to the 20-mm embryos. The corpus ossis pubis, corpus ossis ischii and corpus ossis ilii are present but their volume is considerably less. A thick layer of mesenchyme, which already outlines the volume reached in the 20-mm embryos, surrounds them all. The ramus inferior ossis pubis is not present in cartilage, but as clustered chondrificating cells (Fig. 57). Left and right os pubis do not reach the midline.

A distinct feature of the stage-18 embryos is the apparent lack of histological differentiation of the pelvic content. The pelvis seems to be completely filled with dense and compact mesenchyme. The hindgut and the sinus urogenitalis are clearly outlined and recognizable as are the nerves and blood vessels. However, all the interposing tissue is diffusely arranged and shows clustering or condensation as a sign of the development of other different anatomical pelvic components such as muscles. At this stage of embryonic development, it becomes difficult to denominate the different components around the corpus pelvinum. Exclusion of possibilities and knowledge of the form and position of the muscles in the older embryos and fetuses make identification of the muscles of the pelvic diaphragm possible.

In the WK7435 embryo, no histological clustering of cells was visible in the area of the pelvic diaphragm muscles. This could mean that the histological development of the pelvic diaphragm begins at about embryonic stage 18.

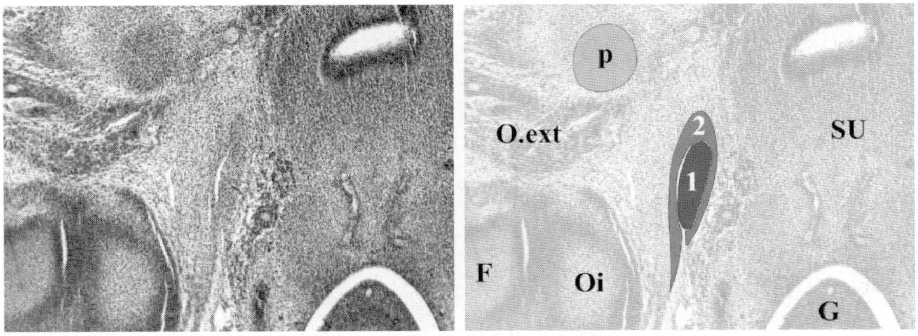

Fig. 57 *F*, femur; *G*, hindgut; *O.ext*, m. obturatorius externus; *Oi*, os ischii; *p*, ramus inferior os pubis; *SU*, sinus urogenitalis; *1*, m. levator ani; *2*, developing fascia. The ramus inferior ossis pubis is outlined in chondrificating cells. No cartilage is present yet. The m. levator ani is formed by clustered mesenchyme, surrounded by a layer of cells representing the developing fascia of the muscle. The fascia stretches out dorsally towards the dorsolateral side of the os ischii

3.4.2 Musculus Obturatorius Internus

The m. obturatorius externus is present on the lateral side of the foramen obturatum. It attaches to the inner margin of the developing ramus inferior ossis pubis, as is illustrated in Fig. 57. The membrana obturatoria has not developed yet. A broad band of mesenchyme fills the pelvic aspect of the foramen obturatum. The m. obturatorius internus is not present at the level of the foramen obturatum.

At the inner level of the ramus ossis ischii and corpus ossis ischii, the m. obturatorius internus is visible (Fig. 58). A large border of developing fascia surrounds it. The muscle's mesenchyme can be seen leaving the pelvis around the dorsal edge of the corpus ossis ischii (Fig. 60). The muscle shows little striation as a sign of early development.

When it appears, the m. obturatorius internus is covered on its medial side by the developing fascia of the m. levator ani. This developing fascia can be recognized as a dense strand of clustered cells running from the m. levator ani towards the corpus ossis ischii, covering the m. obturatorius internus. The m. levator ani itself lies in front of the m. obturatorius internus (Fig. 58). Following the m. obturatorius internus through the embryo, it turns away laterally to eventually leave the pelvis. As the m. obturatorius internus pulls away, the fossa ischiorectalis is present (Fig. 59).

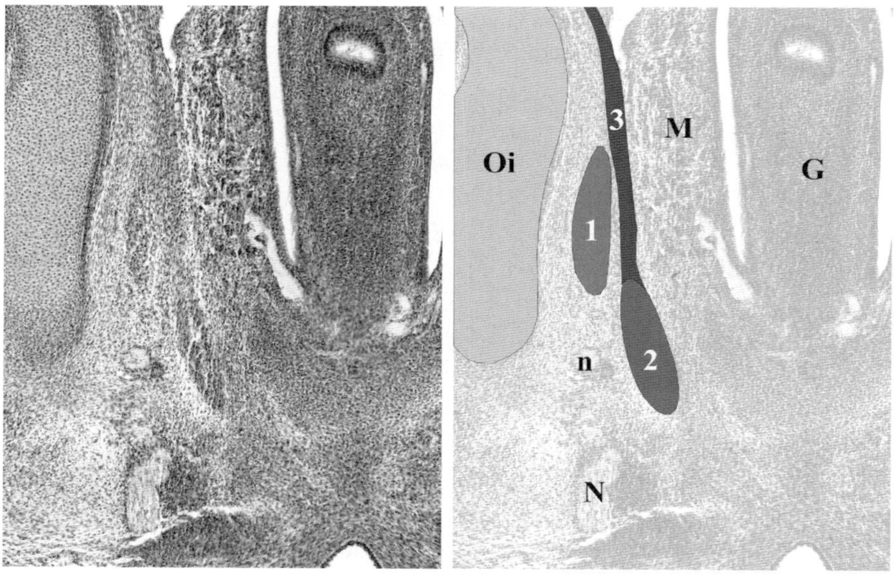

Fig. 58 *G*, hindgut; *M*, embryonic mesothelium of sinus urogenitalis, containing small blood vessels; *N/n*, nervus pudendus and branch; *Oi*, os ischii; *1*, m. obturatorius internus; *2*, m. levator ani; *3*, developing fascia of m. levator ani. The level of the section photographed is just in front of the anal sphincter complex. The level of muscular differentiation of the levator ani has increased compared to the level at the os pubis

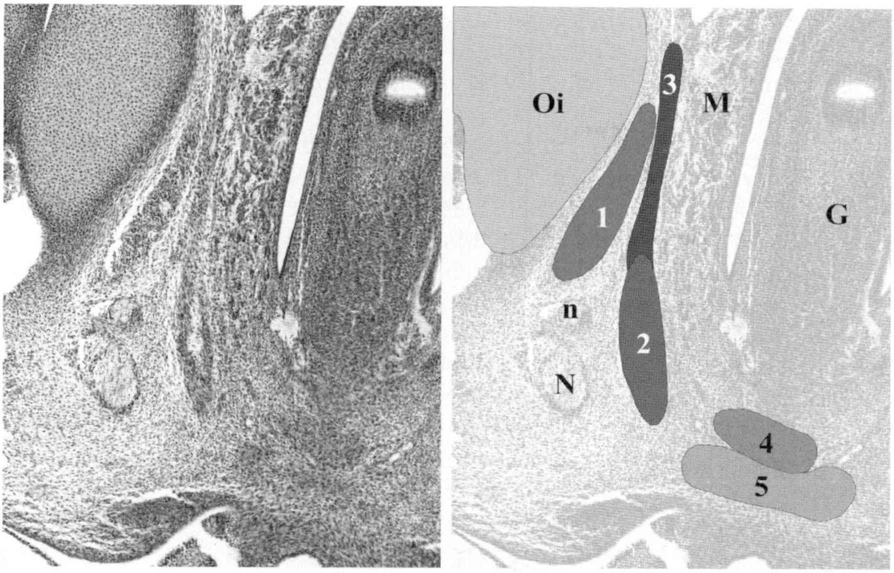

Fig. 59 *G*, hindgut; *M*, embryonic mesothelium of sinus urogenitalis, containing small blood vessels; *N/n*, nervus pudendus and branch; *Oi*, os ischii; *1*, m. obturatorius internus; *2*, m. levator ani; *3*, developing fascia of m. levator ani; *4*, deep part anal sphincter; *5*, superficial part anal sphincter. The levator ani approaches the anal sphincter complex. The fossa ischiorectalis is visible with the nervus pudendus

3.4.3 Musculus Levator Ani

At the lower level of the foramen obturatum, between the sinus urogenitalis and the mesenchymal thickening of the developing membrana obturatoria, a dense clustering of cells appears (Fig. 57). Its ventral margin lies well away from the ramus inferior ossis pubis in a dorsomedial direction. The dorsal margin of the clustered cell group points towards the os ischium. Following it through the embryo it is undoubtedly the developing m. levator ani.

As in the 20-mm embryos, the ventral margin is embedded in undifferentiating mesenchyme that surrounds the sinus urogenitalis laterally and ventrally. At this level near the os pubis and lower edge of the foramen obturatum, the clustered cells do not yet have the histological appearance of muscle cells. With the foreknowledge of it being the m. levator ani, a developing muscle and fascial part is recognized.

The lateral side of the m. levator ani is first covered by mesenchyme running towards the foramen obturatum. A border of numerous small blood vessels and mesenchyme, separating the muscle from the sinus urogenitalis, covers the medial side.

At the level of the ramus ossis ischii, the m. obturatorius internus appears, dorsolateral from the m. levator ani (Fig. 58). At this point, the m. levator ani lies beside the hindgut. It is remarkable that at this level the m. levator ani shows more histological features of muscle cells, i.e., striation. The ventral to dorsal extension of the muscle itself is small. It is extended dorsally by the above-mentioned strand of cells forming the developing fascia of the m. levator ani, overlying the m. obturatorius internus (Fig. 59).

Following the m. levator ani farther in the direction of the os coccygis, the muscle shows more differentiation as the anal sphincter complex is approached. The ventral margin of the m. levator ani is lost in the deep part of the m. sphincter ani externus. This means that the m. levator ani is connected to and has a close relation with the deep external anal sphincter. At this embryonic stage, a clear separation between a superficial and deep part can already be made (Fig. 60). The lateral aspect of the m. levator ani borders the fossa ischiorectalis. Medially the m. levator ani is separated from the hindgut by mesenchyme.

Passing behind the hindgut attachment of the m. levator ani to the anal sphincter complex is lost (Fig. 61). The ventral margin of the levator ani is formed by loose mesenchyme between the anal membrane and the muscle. On its lateral side can

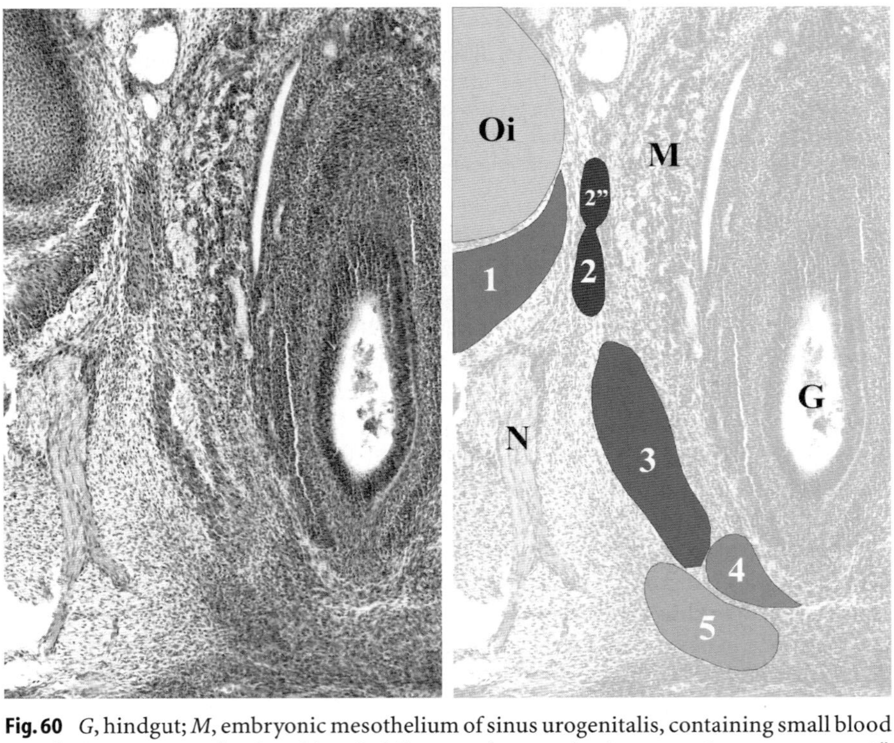

Fig. 60 *G*, hindgut; *M*, embryonic mesothelium of sinus urogenitalis, containing small blood vessels; *N*, nervus pudendus; *Oi*, os ischii; *1*, m. obturatorius internus; *2*, m. coccygeus; *2''*, ligamentum sacrospinale; *3*, m. levator ani with nerve inside; *4*, deep part anal sphincter; *5*, superficial part anal sphincter. The m. obturatorius is leaving the pelvis. The m. levator ani attaches to the deep anal sphincter complex

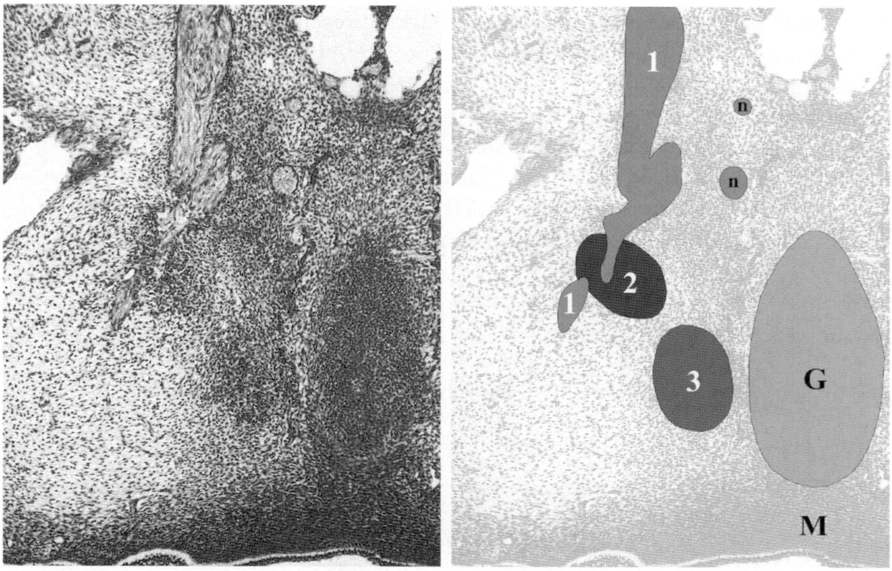

Fig. 61 *G*, hindgut (serosa); *M*, anal membrane with last part of sphincter complex; *n*, small branches of plexus pelvicus; *1*, ramus ventralis S-I continuing in nervus pudendus; *2*, m. coccygeus; *3*, m. levator ani. The level of the section photographed is just behind the hindgut. The m. levator ani lost contact to the sphincter complex and travels towards the os coccygis. Its level of differentiation decreases. The m. coccygeus is pierced by a branch of the nervus pudendus, causing separation in a dorsal and ventral part

be seen tissue of the developing buttocks. Medially, the m. levator ani becomes attached to the developing ligamentum anococcygeum, which presents itself as a dense cluster of mesenchyme running from the hindgut towards the os coccygis. Attachment of the m. levator ani to the os coccygis is not visible; it is lost in the ligamentum anococcygeum before the os coccygis is reached. The m. levator ani shows a decreasing differentiation after it has passed the sphincter complex and is lost in the ligamentum anococcygeum.

3.4.4 Musculus Coccygeus

At the surface of the most dorsal part of the corpus ossis ischii, just above the point where the m. obturatorius internus leaves the pelvis, the m. coccygeus appears (Fig. 60). At its point of origin, no sign of a developing spina ischiadica can be noted, nor can the arcus tendineus levatoris ani be identified in this region.

Following the m. coccygeus towards the os coccygis, its volume increases slightly. The level of muscular differentiation decreases as the m. coccygeus travels towards the os coccygis. Its level of differentiation is higher in the region of the corpus ossis ischii.

A subdivision in a ventral and dorsal part is present, although it is difficult to identify the separate parts in all sections. The subdivision becomes evidently clear when the m. coccygeus is pierced by a nerve (Fig. 61). The individual direction of the ventral and dorsal part can not be identified as the level of differentiation decreases towards the os coccygis and os sacrum.

The distance between the dorsal edge of the m. levator ani and the ventral edge of the m. coccygeus appears to be much smaller than in the older embryos and fetuses. The two muscles almost meet (Figs. 60 and 61).

3.4.5 Musculus Sacrococcygeus Ventralis

The m. sacrococcygeus ventralis cannot be identified in these embryos.

3.5 Stage 17 and Earlier

In the embryonic stages preceding stage 18, no muscles of the pelvic diaphragm could be determined by the presence of (striated) muscle fibers. Clustering of mesenchymal cells in histologically identifiable muscles or fascia is not visible in the HE-stained sections available. The clustering of mesenchymal cells is mandatory to identify the different pelvic structures, since the HE staining cannot distinguish the future histological fate of the different mesenchymal cells. This is a disadvantage of HE staining in the early period of embryological development and makes it impossible to follow the development of the pelvic diaphragm muscles before stage 18.

Nowadays the availability of different antibodies provides the opportunity to distinguish the various developing structures in the mesenchyme. This would make it possible to identify the mesenchymal cells forming the pelvic diaphragm at earlier embryological stages. Ethical issues, however, prevented us from gathering human embryos for antibody staining. The staining methods must focus on developing skeletal muscle cells and on developing fascia.

Striated cells express myoglobin, desmin, alpha-smooth muscle actin, and occasionally S-100, vimentin, and Leu-7. Other antigens such as myosin, creatine kinase, beta-enolase, and titin are less sensitive. Staining the mesenchyme with these markers will identify the mesenchymal cells differentiating into skeletal muscle cells.

Fascia consists of connective tissue cells, forming large amounts of collagen, mainly type I and type III. Type III is produced by fibroblasts before the stronger type I collagen is synthesized. For classic histological staining of collagen, Masson trichrome and methyl violet stains can be used. Combining these staining techniques with an immunohistochemical approach, using anti-collagen type I and type III, will identify mesenchymal cells developing into connective tissue that constitutes the fascia.

We have shown the presence of mixed differentiation of the pelvic diaphragm muscles in the stage-18 embryos. At this stage, embryos are present that do not show any histological clustering of mesenchymal cells. There are also embryos present that show clustering of mesenchymal cells as the earliest histological sign of differentiation. Even within a single embryo, the presence of undifferentiated mesenchyme and early differentiating muscles is shown. This means that at stage 18, the histological development of the pelvic diaphragm begins. We therefore do not have to extend the period of interest before stage 18 to describe the development of the pelvic diaphragm muscles.

Future research, with the use of antibody staining, should focus on the origin of the mesenchymal cells forming the pelvic diaphragm muscles. In the 17-mm to 20-mm CRL stages, we have described the presence of mesenchymal clustering in bands forming either muscle or fascia. This raises a second interesting question for future research. How do the different mesenchymal origins of the muscles and fascias interact? Does mesenchymal clustering of the future fascia precede the formation of the muscle's mesenchyme, or the other way around? It could well be that the formation of fascia from mesenchyme is dependent on the developmental stage of the muscle which it surrounds.

4 Conclusions

This topographical description of the development of the pelvic diaphragm muscles indicates that the start of strong mesenchyme clustering and differentiation into muscle fibers is at stage 18–19 (17 mm CRL). Van der Putte (2005) describes the presence of mesenchyme in the puborectalis area in 13-mm embryos. We could not find this mesenchymal clustering in embryos smaller than 17 mm; we did find the m. levator ani to be more differentiated in the anal sphincter area than in the area towards the os pubis and os coccygis. Mesenchymal clustering present in 13-mm embryos thus makes sense. A fully developed pelvic diaphragm is present in the early fetal period at stage 43–50 mm.

Following the development backwards, differentiation of the mesenchyme into striated muscle fibers within the pelvic diaphragm diminishes and this makes it difficult to identify the pelvic diaphragm muscles in the early embryonic stages. Nevertheless, by knowing the topography of the muscles in the later stages, and carefully examining mesenchymal clustering, the earliest stages could be interpreted.

Differentiation during development has an automatic component (Chandebois and Faber 1983). "In a homogeneous cell population autonomous progression is an automatic process based on a continuous exchange of information (homotypic interactions) which is effected through cell contact relations. The program for this autonomous progression is not communicated to the individual cells of the population at the time of determination. Rather, the program assigns to the cells a certain elementary social behavior, which is the basis for the structuring of the population (Chandebois and Faber 1983).

Mesenchymal clustering will thus automatically produce the striated fibers of the muscles involved. It is this automation, followed backwards, that ensures that the use of topography and mesenchymal clustering effectively demonstrates the pelvic diaphragm muscle development.

One should note that even in 17-mm embryos the anal sphincter complex is easily recognizable and contains muscle fibers. The anal sphincter complex is also the place where the earliest differentiation of muscle fibers of the levator ani is noted. In fact, it is the contact zone between the m. levator ani and the deep part of the m. sphincter ani externus that is recognizable up to 17 mm CRL.

4.1 Musculus Obturatorius Internus

The m. obturatorius internus plays an important role in the structure of the pelvic diaphragm. It covers the lateral walls of the pelvic cavity. Over its inner surface lies the arcus tendineus levatoris ani, which fixes the m. levator ani to the pelvic walls. It also acts as a landmark in the identification of the pelvic diaphragm muscles.

In adults, the m. obturatorius internus is described as a muscle originating from the margins of the foramen obturatum, the medial surface of the os pubis, the medial aspect of the os ischii and the inner surface of the membrana obturatoria (Hafferl 1969). It runs towards the incisura ischiadica minor where it leaves the pelvic cavity through the foramen ischiadicum minus to insert on the inner surface of the trochanter major (fossa trochanterica) as muscle fibers or as a tendon. Nerve branches directly arising from the plexus sacralis innervate the muscle.

Because of the lack of literature on the m. obturatorius internus in humans, and because of its role in the pelvic diaphragm, we have studied the development of this muscle in the previously described human embryos and fetuses.

4.1.1 Summary

The m. obturatorius internus first appears in the pelvic cavity of embryos with a crown rump length of 17 mm, at about 49 days postconception. In this period, the growing embryo should have reached developmental stage 18–19. In younger embryos, only mesenchyme is present, making it impossible to differentiate muscular development with the HE staining of the available embryonic sections.

17 mm (± 49 Days Postconception, Stage 18–19)

At the inner surface of the os ischii at the upper level of the ramus ossis ischii a condensation of cells appears, surrounded by mesenchymal cells. It curves around the dorsal margin of the os ischii, leaving the pelvic cavity at a level just below the future spina ischiadica and runs towards the head of the femur. This was determined as the developing m. obturatorius internus and its developing fascia. The muscle does not show any striation or strands as a sign of early development. The membrana obturatoria has not developed yet. Instead, a broad layer of mesenchyme fills the foramen obturatum. Figure 62 a is a schematic representation of the muscle, showing that its ventral extension does not reach the dorsal margin of the foramen obturatum.

22–23 mm (± 53–54 Days Postconception, Stage 21)

The m. obturatorius internus has extended slightly caudal over the inner aspect of the ramus ossis ischii and slightly ventral reaching the dorsal margin of the foramen obturatum. The dorsal extension towards the femur head remains identical (Fig. 62b). It now begins to show formation of muscle fibers. The membrana obturatoria has not developed yet, but is outlined in mesenchyme.

28–30 mm (± 58–59 Days Postconception, Stage 23)

At the end of the embryonic period, the m. obturatorius internus has started to cover part of the foramen obturatum (Fig. 62c). It has extended mostly in a ventral and caudal direction following the lower margin of the foramen obturatum, i.e., the ramus ossis ischii, but also slightly in a cranial direction towards the corpus ossis ischii. The muscle nearly covers the dorsal half of the foramen obturatum, has clear strands and is covered by a fascia that does not appear to have reached the os pubis. The foramen obturatum is filled by mesenchyme, which starts to condense, forming a more or less recognizable membrana obturatoria.

34–35 mm (± 9 Weeks Postconception)

At the beginning of the fetal period, the m. obturatorius internus has extended farther in a ventral–caudal direction, following the lower margin of the foramen obturatum, and in a ventral–cranial direction, covering the corpus ossis ischii. The muscle now covers about three-quarters of the foramen obturatum, but has not reached the future canalis obturatorius yet (Fig. 62d). The fascia of the m. obturatorius has already reached the os pubis. In the embryos described, the dorsal extension of the muscle from the incisura ischiadica minor towards the femur head consists of fascia and not of muscle fibers. The membrana obturatoria is now clearly recognizable as a small layer of connective tissue between the m. obturatorius internus and externus. At the level of the nervus obturatorius, it has not yet developed into a clear canalis obturatorius.

42–43 mm (± 10 Weeks Postconception)

The m. obturatorius internus has extended farther in a ventral and cranial direction towards the future canalis obturatorius. It covers the foramen obturatum for more than three-quarters of its length, failing to reach the ramus superior ossis pubis (Fig. 62e). The fascia of the muscle is still attached to the os pubis. The membrana obturatoria shows progressing condensation around the nervus obturatorius, beginning to form a canalis obturatorius.

50 mm (± 12 Weeks Postconception)

At 12 weeks postconception, the m. obturatorius internus has nearly reached its full extension, except for a small margin at the ventral caudal edge of the foramen

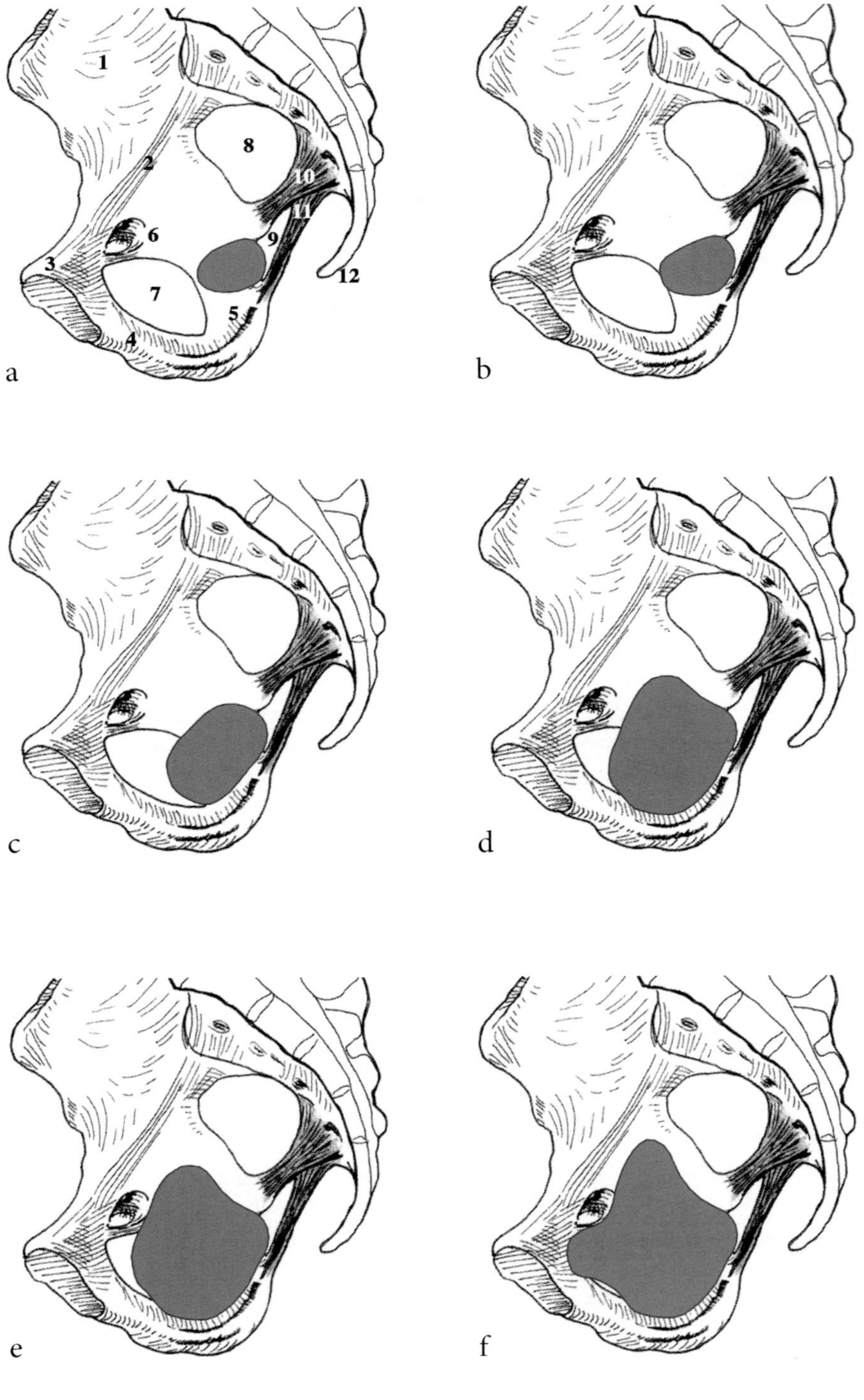
1
2
3
4
5
6
7
8
9
10
11
12
a
b
c
d
e
f

Fig. 62a–f Developing m. obturatorius internus (in *gray*), projected on the mature bony pelvis. *1*, ala ossis ilii; *2*, linea arcuata; *3*, (ramus superior) os pubis; *4*, ramus inferior ossis pubis; *5*, ramus ossis ischii/tuber ischiadicum; *6*, canalis obturatorius; *7*, foramen obturatum; *8*, foramen ischiadicum majus; *9*, foramen ischiadicum minus; *10*, ligamentum sacrospinale; *11*, ligamentum sacrotuberale; *12*, os coccygis

obturatum. It has extended farther in a ventral and cranial direction, surrounding the lower margin of the canalis obturatorius and almost reaching the incisura ischiadica major and linea arcuata. It has extended ventrally to insert on the os pubis (Fig. 62f). The canalis obturatorius has now developed completely. The dorsal extension towards the femur head consists of muscle fibers.

4.1.2 Conclusion

One very important conclusion can be drawn from the described development of the m. obturatorius internus, namely that the muscle develops in a dorsal to ventral direction. We were not able to follow the development of the muscle in embryos younger than 17 mm CRL, but the fetal results of the dorsal to ventral development make us conclude that the margins of the foramen obturatum and the caudal edge of the foramen ischiadicum majus are not the origins of the m. obturatorius internus, but the insertions.

Using specific antibodies, it would be possible to follow the development of the m. obturatorius internus in younger embryos, to find the origin of the muscle. In the embryos described, we have seen that the extrapelvic part of the m. obturatorius internus (from incisura ischiadica minor towards trochanter major femoris) can be muscle, fascia, and/or tendon. This is in accordance with the adult situation as described above. We have also shown that the fascia of the m. obturatorius internus precedes the developing muscle, as it reaches the os pubis before the muscle fibers do. It might therefore be possible that the origin of the fascia and the muscle fibers of the m. obturatorius internus is situated at the trochanter major femoris, or at the level of the incisura ischiadica minor, or both.

The development of the membrana obturatoria is interesting. It shows a developmental trend similar to that of the m. obturatorius internus. In other words, the development of the membrana obturatoria precedes that of the m. obturatorius internus in a ventral direction.

The development of the membrana obturatoria seems to be induced by the m. obturatorius internus. Although the m. obturatorius externus is already present before the obturatorius internus reaches the other side of the membrana obturatoria, the area of the membrana remains mesenchyme-like. The farther the m. obturatorius internus extends over the membrana obturatoria, the more typical the connective tissue appearance of the membrana. This induction or heterotypic action follows certain rules: “The inducer re-initiates or deflects the course of autonomous progression but it does not establish its program. It selects one program from among others, which are already being executed. The execution of the program—which under certain experimental conditions may continue or be accelerated in the absence of the inducer—consists in the stimulation of certain luxury metabolic strategies. The cells become competent to respond to the action of the inducer as soon as the inductive stimulus suffices to raise those strategies above a threshold beyond which they manifest themselves in visible structure (Chandebois and Faber 1983)". The autonomic progression of the mesenchymal area at the future membrana obturatoria is promoted to differentiate into connective tissue, with the m. obturatorius internus as inducer.

4.2 Musculus Levator Ani

The m. levator ani is the main muscle that constitutes the pelvic diaphragm. Difference of opinion exists on the presence of a subdivision into distinct muscles. The m. levator ani is described as a single muscle, subsequently divided into an imaginary pubococcygeal and iliococcygeal part and as a group of muscles (see Sect. 1.2.2). This group of muscles then consists of the m. pubococcygeus and the m. iliococcygeus. Some authors describe the presence of a third muscle, the m. puborectalis. Even the m. coccygeus (m. ischiococcygeus) is included in the levator ani by some authors.

Little is known about the development of the m. levator ani. The development during the embryonic period is only described by Popowsky (1899) and Power (1948). Popowsky concluded that the m. levator ani developed from the m. coccygeus by ventral migration of part of its muscle fibers. Power described the m. levator ani as part of the m. rectus abdominis, separated by ingrowth of the pubic bone. Fritsch (1994) discussed the development of the m. levator ani in the fetal period, although she only made observations on the anatomical presence and extension of the m. levator ani in different fetal stages.

No description of the actual embryonic and fetal development of the m. levator ani, progressing over time, could be found. We therefore have studied and described the embryological and early fetal development of the m. levator ani and defined it over time.

4.2.1 Summary

The first appearance of a developing m. levator ani is found in embryos with a CRL of 17 mm. In younger embryos, only mesenchyme is present, making it impossible to differentiate muscular development with the HE staining of the available embryonic sections.

17 mm (± 49 Days Postconception, Stage 18–19)

The m. levator ani appears as a single muscle at the level of the external anal sphincter complex. It is attached to the deep part of the m. sphincter ani externus and radiates around the hindgut and sinus urogenitalis towards the os pubis and the os coccygis. The m. levator ani is not attached to the sinus urogenitalis. Extending ventrally towards the os pubis, the muscular differentiation of the mesenchyme decreases. The m. levator ani ends as mesenchyme medially and posteriorly of the developing os pubis. Extending dorsally towards the os coccygis, its level of muscular differentiation also decreases. This decrease in differentiation lets us conclude that the m. levator ani starts to develop as a part of the external anal sphincter complex and then extends towards the os pubis and os coccygis, radiating around the hindgut and sinus urogenitalis. Figure 63 a shows a schematic representation of the m. levator ani in an artistic impression of the developing bony pelvis. Ventrally, the left and right os pubis are wide apart. The ventral parts of the left and right m. levator ani end between the two pubic bones, without attachment to these bones. Radiating around the rectum and the sinus urogenitalis, the m. levator ani runs towards the midline, eventually ending in front of the os coccygis at the level of the developing ligamentum anococcygeum. The lateral extension of the developing muscle fibers is limited to a small band. The muscle is surrounded by its developing fascia, which can already be seen extending over the m. obturatorius internus towards the future arcus tendineus levatoris ani.

22–23 mm (± 53–54 Days Postconception, Stage 21)

The m. levator ani now completely consists of fully differentiated muscle fibers. Ventrally it is not yet attached to the developing os pubis. The fascia of the muscle, however, is attached to the developing endopelvic fascia, connecting it to the ramus inferior ossis pubis. The left and right os pubis have extended towards the midline, now reaching in front of the ventral part of the m. levator ani. The ventral to dorsal extension of the m. levator ani has not changed. Dorsally, the m. levator ani extends towards the midline and is attached to the developing ligamentum anococcygeum. At this stage of development, the m. levator ani starts to extend laterally and cranially at a level in between the hindgut and the sinus urogenitalis (Fig. 63b). The m. levator ani is growing towards the ventral part of the arcus tendineus levatoris ani, following its preceding fascia. The muscle has not shown any progression towards the midline.

28–30 mm (± 58–59 Days Postconception, Stage 23)

Reaching the end of the embryonic period, the m. levator ani has shown acceleration in its development. The left and right os pubis have reached the midline, forming the symphysis pubica. With this outgrowth of the os pubis and their contact as the symphysis pubica, the m. levator ani has extended medially at its ventral margin. It is now more clearly enclosing the sinus urogenitalis. The outgrowth of the os pubis towards the midline and the attachment of the m. levator ani to the os pubis create the hiatus urogenitalis.

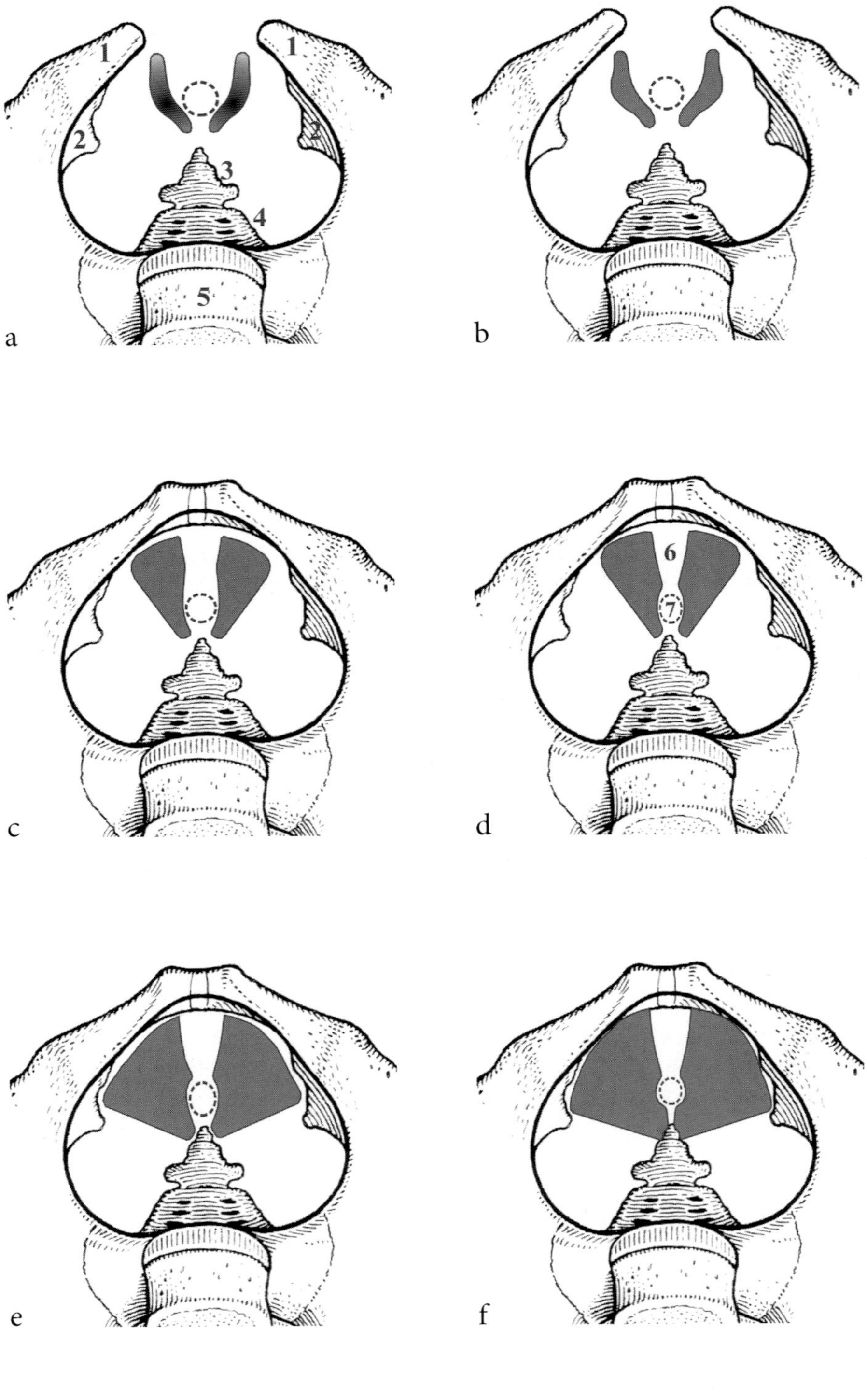
1
1
2
2
3
4
5
a
b
c
6
7
d
e
f

◄

Fig. 63a–f Developing m. levator ani (in *gray*) projected on the mature bony pelvis. An artistic impression has been made for the pelvis, combining the adult bony pelvis with developing structures. In Fig. 67, the muscular differentiation is represented in *gray shadings with black* representing differentiated muscle fibers and *light gray* representing mesenchyme. *1*, os pubis; *2*, developing spina ischiadica; *3*, os coccygis; *4*, os sacrum; *5*, fifth lumbar vertebra; *6*, hiatus urogenitalis; *7*, hiatus ani; *dotted line*, rectum/anus outline

The levator ani is not attached to the sinus urogenitalis. At its ventral margin, the m. levator ani has also extended cranially and dorsally, following the arcus tendineus levatoris ani. Projecting over the foramen obturatum, it is now covering the ventral half of the foramen obturatum and the m. obturatorius internus (Fig. 63c). The dorsal extension of the m. levator ani has slightly progressed towards the tip of the os coccygis. It is still attached to the ligamentum anococcygeum and not to the os coccygis itself.

34–35 mm (± 9 Weeks Postconception)

At the beginning of the fetal period, the m. levator ani has shown little progression (Fig. 63d). It has not extended farther dorsally towards the os coccygis nor has it extended dorsally and cranially, following the arcus tendineus levatoris ani. The ventral margin of the levator ani, however, has extended farther towards the inner aspect of the ramus inferior ossis pubis. With the progressing development of the endopelvic fascia, the distance between the os pubis and the m. levator ani has also decreased. Between the sinus urogenitalis and the rectum, the m. levator ani has extended towards the midline. The formation of a separate hiatus urogenitalis and hiatus ani now becomes visible. There is no attachment of the m. levator ani to the sinus urogenitalis noted.

42–43 mm (± 10 Weeks Postconception)

One week later in its development, the m. levator ani has again shown substantial progression. Towards the os coccygis it has extended farther, now almost reaching the tip of the os coccygis. The left and right mm. levatores ani are projected near the lateral surface of the tip of the os coccygis. Attachment is made through the ligamentum anococcygeum. The most prominent progression is reached in the dorsal and cranial direction, following the arcus tendineus levatoris ani. The m. levator ani now almost reaches the dorsal end of the arcus tendineus levatoris ani, near the developing spina ischiadica where the ATLA attaches (Fig. 63e). The m. levator ani has shown no progression towards the midline; the future hiatus urogenitalis and hiatus ani are the same as 9 weeks postconception.

50 mm (± 12 Weeks Postconception)

The m. levator ani has shown a marked progression towards the midline, especially between the sinus urogenitalis and the rectum, at the level of the future centrum tendineum and behind the rectum at the level of the ligamentum anococcygeum.

With the continuing development of the ligamentum anococcygeum, the left and right musculi levatores ani are almost attached to each other, only separated by the ligamentum anococcygeum. The levator ani is now attached to the upper aspect of the ramus inferior ossis pubis on the ventral side. On the lateral side, the m. levator ani covers the pelvic outlet completely, following the arcus tendineus levatoris ani towards the future spina ischiadica. At 12 weeks postconception, the m. levator ani has almost developed completely (Fig. 63f). The adult situation is not yet reached at the level of the future centrum tendineum only.

4.2.2 Conclusions

Our developmental study presented shows that the m. levator ani is a single muscle. There is no evidence of an anatomical subdivision in an m. pubococcygeus, m. iliococcygeus, or m. puborectalis. In particular, the existence of a m. puborectalis or puborectal sling is refuted by the results presented. No muscular fibers actually cross the midline, so the formation of a muscular sling around the rectum is improbable.

The m. levator ani develops as a part of the external anal sphincter complex and is attached to the deep part of the m. sphincter ani externus. It extends ventrally towards the os pubis encircling the sinus urogenitalis and dorsally towards the os coccygis. When the m. levator ani reaches the level of the os pubis, which is also caused by the ingrowth of the pubic bone (the hiatus urogenitalis is created), it starts to extend over the pelvic outlet.

This extension is not constant over time. Particularly at the end of the embryonic period, when the symphysis pubica is formed, and between the 9th and 10th week postconception, accelerated development is noted. The first acceleration is caused by the growing pubic bone, which more or less drags the m. levator ani towards the midline, outlining the hiatus urogenitalis. This ingrowth of the os pubis towards the midline is probably the basis for Powers's (1948) view on the development of the pelvic diaphragm. The reason for the second acceleration is not clear.

After it reaches the pubic bone, the m. levator ani extends dorsally and cranially over the pelvic outlet, following its developing fascia and the developing arcus tendineus levatoris ani. In contrast to Popowsky (1899), who described a ventral migration of muscle fibers, our study describes a dorsal migration. The left and right musculi levatores ani fan out over the pelvic outlet like a peacock spreading its tail feathers.

The extension of the m. levator ani towards the midline is mainly caused by external factors. As already described, the ventral medial extension is caused by the developing os pubis. The medial extension, between the sinus urogenitalis and the rectum, is caused by the developing centrum tendineum (perineal body), pulling the left and right levator ani towards each other (see the extensive research of van der Putte [2005] for further details on the development of the human perineum). Behind the rectum, the medial closure of the pelvic outlet by the left and right levator ani is mediated by the developing ligamentum anococcygeum. The ligamentum anococcygeum reduces from a broad mesenchymal band to a thin band of connective tissue, connecting the deep part of the m. sphincter ani externus and the left and right m. levator ani with the tip of the os coccygis.

The observations on the muscles of the pelvic diaphragm by Fritsch and Fröhlich (1994) are not shared. The existence of separate musculi pubococcygeus, iliococcygeus, and puborectalis cannot be reproduced. Their description of the levator ani originating from the pelvic surface of the os pubis is probably confounded by misinterpretation of the varying muscular differentiation of the m. levator ani at the 17-mm stage, as noted earlier. The differences between male and female pelvic diaphragm, described by Fritsch, are not found in our study. The pelvic diaphragm develops independently of the sinus urogenitalis and therefore a sex-dependent difference in the composition of the pelvic diaphragm makes no sense. However, development of the urogenital diaphragm and perineum is sex-dependent, as van der Putte (2005) demonstrated.

The development of the pelvic diaphragm is independent of the developing sinus urogenitalis and only indirectly dependent on the development of the rectum. The pelvic mediastinum described by Marani (2002) can only be partly reproduced in this study. The sinus urogenitalis is not anywhere directly attached to the m. levator ani. The anchoring of the sinus urogenitalis to the pelvic diaphragm is made possible through the urogenital diaphragm. The m. levator ani is not attached to the rectum but to the deep part of the external sphincter. This connection to the deep part of the external sphincter is responsible for the functional combined actions of the rectum and pelvic diaphragm. A complete free space for the individual free development of the pelvic organs, as described by Marani (2002), is thus only present for the urogenital organs and not for the anal sphincter area.

The projection of the pelvic outlet in the embryonic and fetal stages is to the front. Because of this projection, the m. levator ani extends more dorsally than cranially during its development. In later fetal life, with the formation of the lumbar lordosis, the formation of the promontorium and the backward stretching of the os coccygis, the pelvic outlet will rotate to its final position in a dorsal caudal projection. At that time, the os pubis will be the lowest part of the bony pelvis and not the os coccygis (Marani 2002). The m. levator ani in the adult dorsal caudal projection would have shown extension in a more cranial direction.

4.3 Musculus Coccygeus

In adults, the m. coccygeus originates from the spina ischiadica and inserts on the lateral inner surface of the os sacrum and os coccygis. With the disappearance of a tail in humans, its function has shifted from motion to support (Thompson 1899). The m. coccygeus can therefore be considered part of the pelvic diaphragm. This shift to a solely supportive function is responsible for the construction of the muscle in adults. It is variably described as consisting of fascia or a combination of muscle fibers and fascia. In adults it rarely consists of only muscle fibers (Thompson, 1899). The m. coccygeus is attached to, or forming part of, the ligamentum sacrospinale.

Apart from Popowsky's work (1899), no literature on the m. coccygeus as part of the developing pelvic diaphragm could be found. Popowsky describes the m. coccygeus as the origin of the developing m. levator ani. The ventral fibers of the m. coccygeus migrate to the front, forming the m. levator ani. The development of the m. coccygeus itself is not described by Popowsky.

4.3.1 Summary

The first appearance of a developing m. coccygeus is found in embryos with a CRL of 17 mm. In younger embryos, only mesenchyme is present, making it impossible to differentiate muscular development with the HE staining of the available embryonic sections.

17 mm (± 49 Days Postconception, Stage 18–19)
Directly above the level where the m. obturatorius internus leaves the pelvis, the m. coccygeus appears. At this stage no developing spina ischiadica is noted. As it appears, the m. coccygeus shows striation as a sign of differentiation into muscle. This differentiation is lost following the m. coccygeus towards the os coccygis, where it consists of mesenchyme. A closer study on the outline of the muscle reveals that it consists of a dorsal and a ventral part, both comprised of muscular tissue. Towards the os coccygis, the m. coccygeus slightly enlarges in a ventral-dorsal dimension. It is lost in mesenchyme well before the os coccygis is reached. Figure 64 a shows an impression of the m. coccygeus, projected in a schematic representation of the developing bony pelvis. It has to be understood that the future spinae ischiadicae have been drawn in to facilitate a better understanding of the actual position of the m. coccygeus, while in fact they do not yet exist.

22–23 mm (± 53–54 Days Postconception, Stage 21)
The m. coccygeus now shows complete striation. It originates at the level of the future spina ischiadica. The ventral-dorsal dimension of the lateral half of the muscle has increased (Fig. 64b). The medial half has extended towards the os coccygis, but without progression in dimension. It is lost in connective tissue

Fig. 64a–f Developing m. coccygeus (in *gray*) projected on the mature bony pelvis. An artistic impression has been made for the pelvis, combining the adult bony pelvis with developing structures. In Fig. 73, the muscular differentiation is represented in *gray shadings with black* representing differentiated muscle fibers and *light gray* representing mesenchyme. The combined lateral to medial and ventral to dorsal progression of the m. coccygeus over time becomes clear. The separation in a ventral and dorsal part of the m. coccygeus is indicated by a *white line*. *1*, os pubis; *2*, developing spina ischiadica; *3*, os coccygis; *4*, os sacrum; *5*, fifth lumbar vertebra

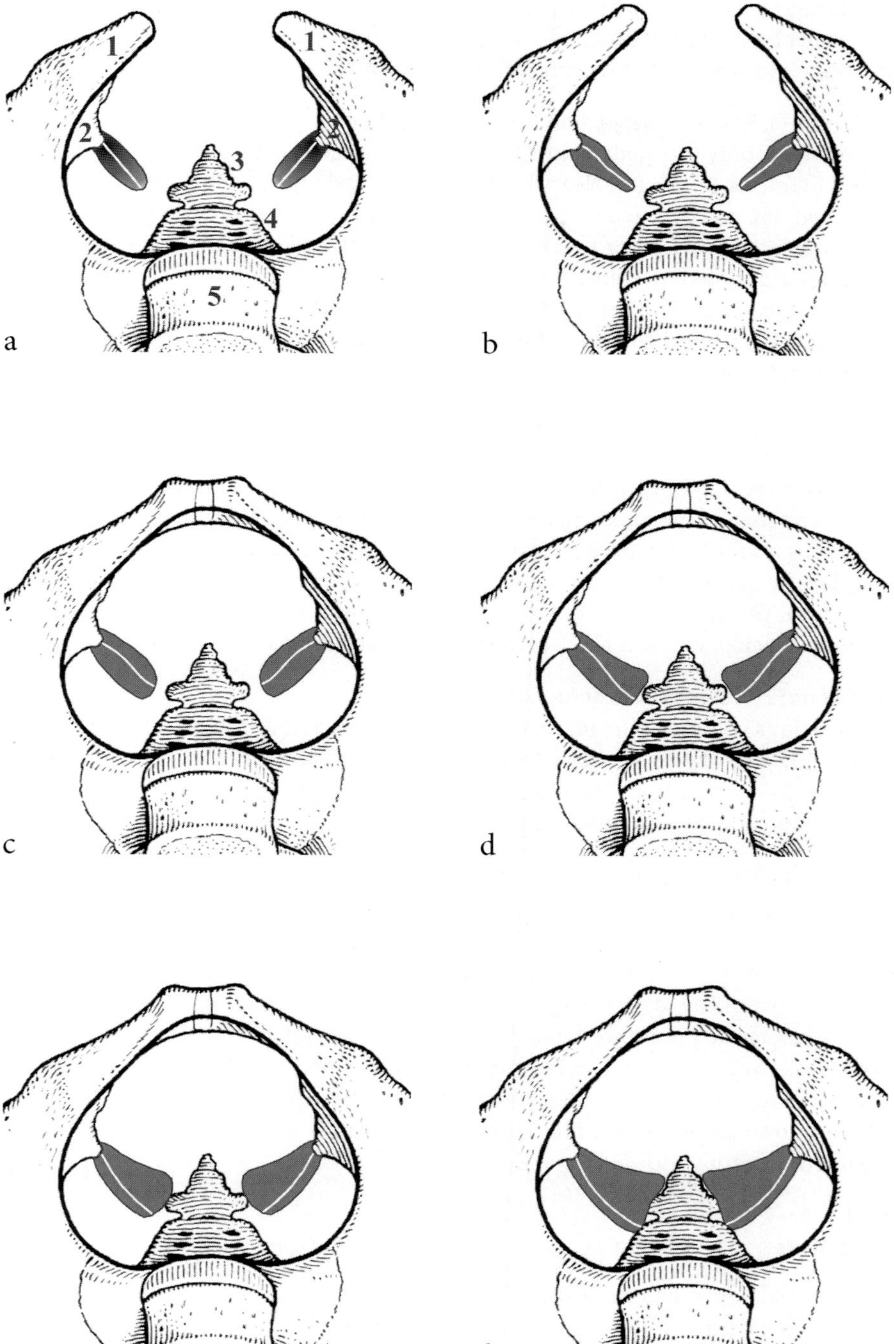
1
1
2
2
3
4
5
a
b
c
d
e
f

extending up to the lateral aspect of the os coccygis and os sacrum. The m. coccygeus has a ventral and a dorsal part. The dorsal part has the largest volume and travels towards the os sacrum, the smaller ventral part towards the os coccygis.

28–30 mm (± 58–59 Days Postconception, Stage 23)

At this stage, only the medial part near the os coccygis has extended in the ventral dorsal diameter (Fig. 64c). Near the future spina ischiadica, which still cannot be identified, the m. coccygeus has shown no progression. The m. coccygeus has also not progressed any farther towards the os coccygis and os sacrum. A dorsal and ventral part is noted. In contrast to the earlier stages, the ventral part has increased in size, but remains smaller than the dorsal part. The ventral part travels towards the lateral aspect of the os coccygis and the dorsal part towards the lateral aspect of the os sacrum.

34–35 mm (± 9 Weeks Postconception)

The m. coccygeus has now extended farther towards the os coccygis and os sacrum. Its overall ventral–dorsal dimension has extended slightly (Fig. 64d). The most important change has taken place between the ventral and dorsal part of the m. coccygeus. At the lateral part, near the future spina ischiadica, the dorsal part of the muscle is larger than the ventral part. At the medial part, towards the os coccygis, the ventral part becomes larger than the dorsal part. The future spina ischiadica starts to develop, and is now discernible as a mesenchymal condensation.

42–43 mm (± 10 Weeks Postconception)

The spacious volume of the m. coccygeus has hardly changed. It has only extended farther towards the os coccygis (Fig. 64e). The ventral part of the m. coccygeus, however, has now become the largest part of the m. coccygeus. The dorsal part has changed into a narrow band of muscle fibers.

50 mm (± 12 Weeks Postconception)

The m. coccygeus has almost reached the margins present in the adult situation, except for its insertion on the os coccygis and os sacrum (Fig. 64f). The developing spina ischiadica is outlined in developing cartilage. The m. coccygeus clearly originates from this developing spina ischiadica. The ventral part of the m. coccygeus has fanned out and is much larger than the dorsal part. The dorsal part can be followed all the way to the lateral surface of the os sacrum, where it is lost in connective tissue inserting on the os sacrum. The dorsal part always remains attached to the ventral part. The ventral part travels towards the lateral surface of the os coccygis.

4.3.2 Conclusions

The developmental characteristics of the m. coccygeus indicate that the origin of the muscle is at the level of the future spina ischiadica, and its insertion on the lateral pelvic surface of the last sacral and first coccygeal vertebrae.

The described development of the m. coccygeus clearly demonstrates the reason for the close relationship of the m. coccygeus and the ligamentum sacrospinale in the adult topographical anatomy. The m. coccygeus develops in a ventral and dorsal part, always connected to each other. The ventral part increases in size during the embryonic and early fetal development and will form the actual m. coccygeus that is described in adult anatomy. With the progressing volume of the ventral part, the dorsal part decreases in volume and will form the ligamentum sacrospinale. In the described developmental stages, the m. coccygeus and ligamentum sacrospinale are both constituted of muscle fibers. In the later fetal stages, or perhaps even after birth, the m. coccygeus and most certainly the ligamentum sacrospinale will have to degrade to fibrous tissue to explain the situation described in adult topographical anatomy. Since the spinae ischiadica, for instance, reach their complete development in late childhood (Marani 2002), it could well be that the m. coccygeus and ligamentum sacrospinale retain their muscular form up until that time. This can be an interesting subject for further research. Modern magnetic resonance imaging could be helpful in studying the composition of the m. coccygeus and ligamentum sacrospinale in children. From an embryological point of view, it can already be concluded that the m. coccygeus and ligamentum sacrospinale are constructed to perform the function of moving the os coccygis. With the evolutionary disappearance of the tail in humans, they will eventually maintain a supportive function and transform to fascia: use it or loose it.

The division of the m. coccygeus into a ventral and dorsal part also explains why Popowsky (1899) believed that the m. levatoris ani developed out of ventrally migrating muscle fibers from the m. coccygeus. In the late embryonic period he studied, the ventral part of the m. coccygeus is much smaller than the dorsal part, as is shown in our study. Popowsky perhaps made his conclusion by taking the dorsal part as the actual m. coccygeus, while it is in fact the ligamentum sacrospinale. The small ventral part, extending from a small band to a large muscle at the disadvantage of the dorsal part, could therefore have been mistaken for the developing m. levatoris ani.

The m. coccygeus is innervated by a nerve branch from the nervus pudendus. This nerve branch pierces in between the dorsal and ventral part of the muscle. Hafferl (1969) describes this innervating nerve branch lying on the pelvic aspect of the m. coccygeus. It is now shown that the branch can also be situated on the extrapelvic aspect of the m. coccygeus.

The projection of the pelvic outlet in the embryonic and fetal stages is to the front. Because of this projection, the m. coccygeus and future ligamentum sacrospinale are projected in a cranial–caudal direction. In later fetal life, with the formation of the lumbar lordosis, the formation of the promontorium, and the backward stretching of the os coccygis, the pelvic outlet will rotate to its final position in a dorsal–caudal direction. With this rotation, the m. coccygeus and future ligamentum sacrospinale will project in a ventral to mediodorsal direction.

4.4 Clinical Implications

The early development of the human pelvic diaphragm muscles has implications for our understanding of pathophysiological conditions of the pelvic diaphragm. We give two examples:

1. The exstrophy–epispadias complex (EEC) describes a congenital malformation, with the urethra opening on the dorsum of the penis (epispadia) in conjunction with classic bladder exstrophy or cloacal exstrophy. In the spectrum of this complex, a complete epispadia with bladder exstrophy is the most common variant, and is also found in females. It is associated with pubic diastasis and lateral displacement of perineal and pelvic floor muscular components, which are underdeveloped. This displacement plays a key role in the cause of urinary and fecal incontinence associated with EEC. Treatment of incontinence in EEC patients focuses on restoration of the displacement by approximation of the lateralized perineal and pelvic floor musculature to the midline, including the pubic diastasis (de Jong 2001; De Gennaro and Mosiello 2005).
   We have shown that the outgrowth of the os pubis towards the midline plays an important role in the formation of the pelvic diaphragm, particularly the hiatus urogenitalis. The midline approximation of the anterior pelvic diaphragm is caused by the outgrowing os pubis, and later in fetal life also by the development of the perineal body. So, embryologically, the wish to approximate the lateralized anterior pelvic floor muscles in the treatment of EEC patients makes sense.
2. Imperforate anus is the absence of a normal anal opening. In low lesions, the anus did not develop and the rectum is covered by skin. In high lesions the rectum ends in a blind pouch, usually connected to the vagina (females) or urethra (males) by a fistula. Treatment of the low lesions consists of a perineal anoplasty. The high lesions are treated by recreating a perineal opening and pulling the rectum through the newly created anus. Although much attention is given to the anal sphincter complex and levator ani, most of these treatments result in fecal incontinence (Abrams et al. 2005; De Gennaro and Mosiello 2005).
   We have shown that the levator ani starts to develop as part of the anal sphincter complex and is attached to the deep part of the m sphincter ani externus. Recreating the anus, and pulling through of the rectum, will cause damage to the close relationship of the levator ani and the anal sphincter complex, together with neurological damage, causing failure in the physiological interaction of the two muscle groups and therefore malfunction in defecation. Recreating the anal sphincter complex by using the m. puborectalis is bound to fail, because the m. puborectalis does not in fact exist, as we have shown.

5
Full Color Illustrations

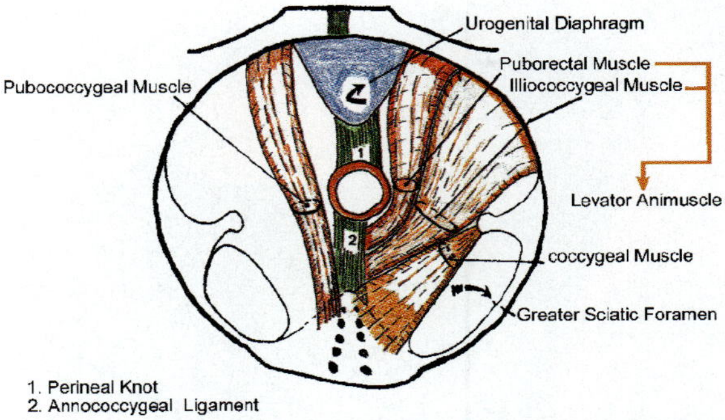

Fig. 1 Muscles of the pelvic diaphragm, as separate muscles. (From Marani 2002, with permission)

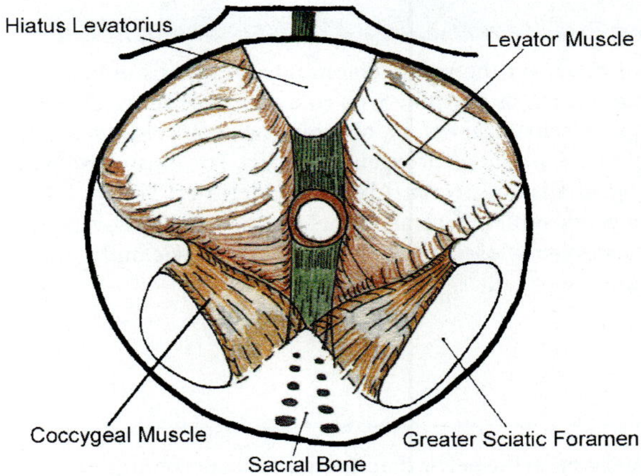

Fig. 2 Muscles of the pelvic diaphragm, as a single muscle. (From Marani 2002, with permission)

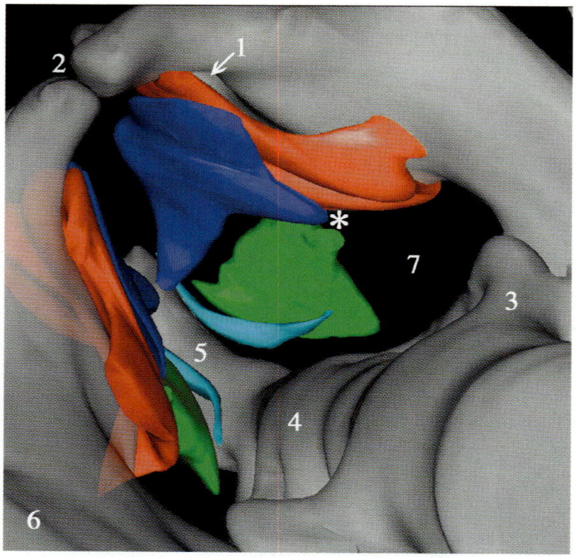

Fig. 12 View from above on the pelvic diaphragm. *1*, canalis obturatorius; *2*, symphysis pubica; *3*, pars lateralis ; *4*, os sacrum; *5*, os coccygis; *6*, os ilium; *7*, foramen ischiadicum majus; *, future spina ischiadica; *red*, m. obturatorius internus; *blue*, m. levator ani; *green*, m. coccygeus; *light blue*, m. sacrococcygeus ventralis. For illustrative purposes, the pelvic viscera (sinus urogenitalis and rectum) have been left out. The m. levator ani closes the ventral half of the pelvic outlet. Ventral to dorsal, it completely covers the m. obturatorius internus. The m. coccygeus closes the dorsal half of the pelvic outlet and has the same size as the m. levator ani

Fig. 13 (top) Frontal view of the pelvic diaphragm. *1*, canalis obturatorius; *2*, m. obturatorius internus leaving pelvis to insert on femur; *3*, acetabulum; *4*, os coccygis; *5*, ramus ossis ischii; *SU*, sinus urogenitalis; *R*, rectum; *red*, m. obturatorius internus; *blue*, m. levator ani; *green*, m. coccygeus; *light blue*, m. sacrococcygeus ventralis. The m. obturatorius internus completely covers the foramen obturatum except for the canalis obturatorius. Note the lateral position of the m. coccygeus in regard to the m. levator ani. The left and right m. levator ani show curving around the sinus urogenitalis and rectum nearing each other at the level of the future centrum tendineum (*white square*) and ligamentum anococcygeum (*white dot*)

Fig. 14 (bottom) View of the pelvic diaphragm from below. *1*, acetabulum; *2*, ramus ossis ischii; *3*, os coccygis; *4*, last sacral vertebra; *5*, os ilium; *SU*, sinus urogenitalis; *R*, rectum; *red*, m. obturatorius internus; *blue*, m. levator ani; *green*, m. coccygeus; *light blue*, m. sacrococcygeus ventralis. Note the four-quadrant formation of the pelvic diaphragm

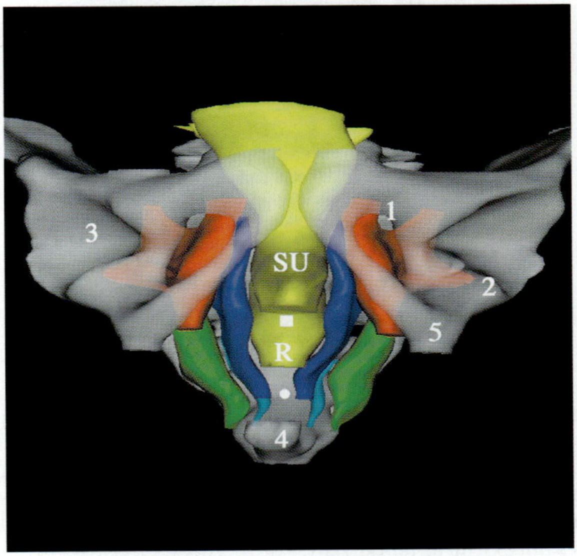
1
3
SU
2
5
R
4

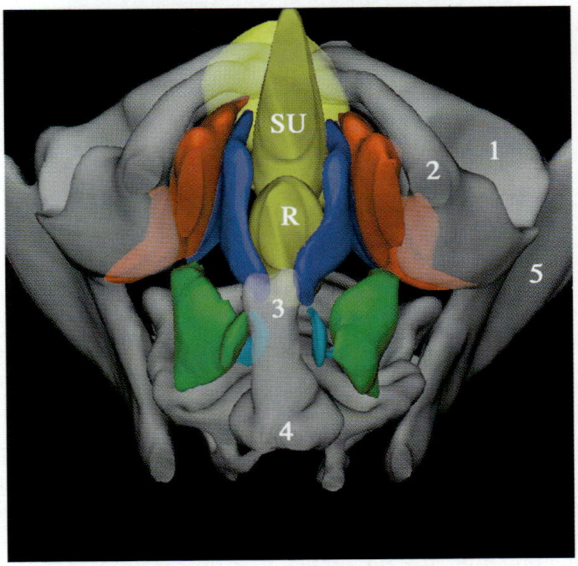
SU
1
2
R
5
3
4

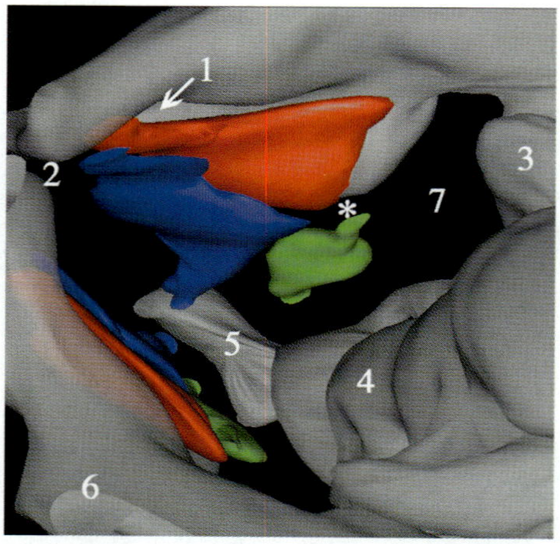

Fig. 21 View from above on the pelvic diaphragm. *1*, canalis obturatorius; *2*, symphysis pubica; *3*, articulatio sacroiliaca; *4*, os sacrum; *5*, os coccygis; *6*, os ilium; *7*, foramen ischiadicum majus; *, future spina ischiadica; *red*, m. obturatorius internus; *blue*, m. levator ani; *green*, m. coccygeus. For illustrative purposes, the pelvic viscera (sinus urogenitalis and rectum) have been left out. The m. levator ani closes the ventral half of the pelvic outlet and almost completely covers the m. obturatorius internus. Compared to the 50-mm fetus, the m. levator ani does not extend as far dorsally. The m. coccygeus closes the dorsal half of the pelvic outlet

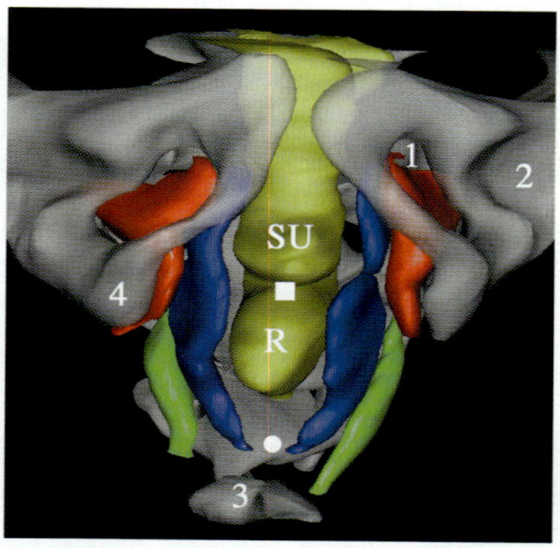

Fig. 22 Frontal view of the pelvic diaphragm. *R*, rectum; *SU*, sinus urogenitalis; *1*, canalis obturatorius; *2*, acetabulum; *3*, os coccygis; *4*, ramus ossis ischii; *white square*, future centrum tendineum; *white dot*, future ligamentum anococcygeum; *red*, m. obturatorius internus; *blue*, m. levator ani; *green*, m. coccygeus. The m. obturatorius internus does not completely cover the foramen obturatum. It is not present near the canalis obturatorius. The left and right mm. levatores ani show curving around the sinus urogenitalis and rectum at the level of the future centrum tendineum and ligamentum anococcygeum. Extension towards the midline is not as advanced as it is in the 50-mm fetus

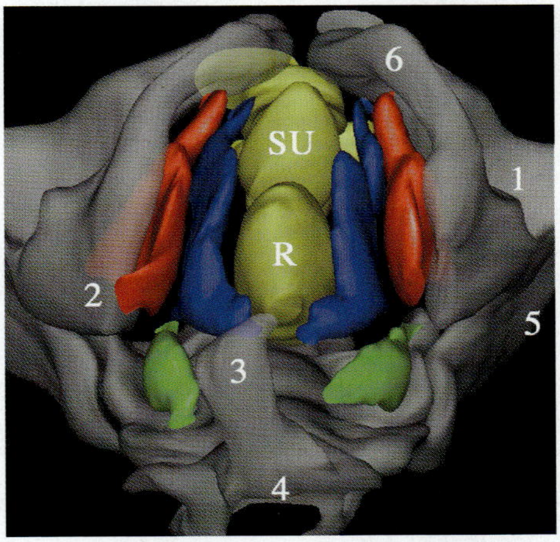

Fig. 23 View of the pelvic diaphragm from below. *R*, rectum; *SU*, sinus urogenitalis; *1*, acetabulum; *2*, ramus ossis ischii/tuber ischiadicum; *3*, os coccygis; *4*, last sacral vertebra; *5*, os ilium; *6*, ramus inferior ossis pubis; *red*, m. obturatorius internus; *blue*, m. levator ani; *green*, m. coccygeus. Note the four-quadrant formation of the pelvic diaphragm. The m. coccygeus is less developed than in the 50-mm fetus

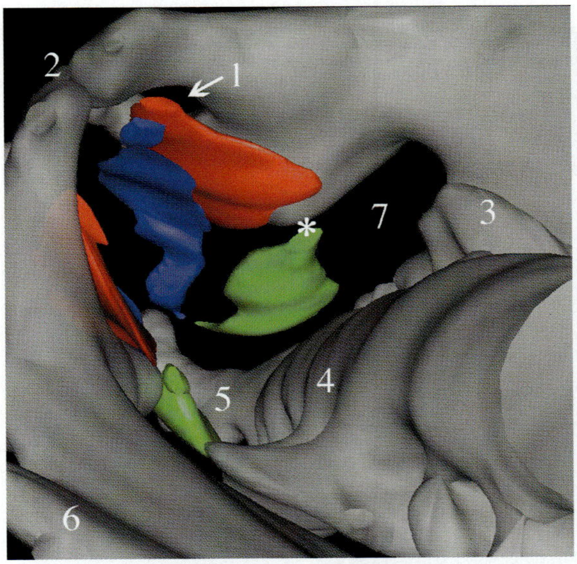

Fig. 30 View from above on the pelvic diaphragm. *1*, foramen obturatum; *2*, symphysis pubica; *3*, pars lateralis os sacrum; *4*, os sacrum; *5*, os coccygis; *6*, ala ossis ilii; *7*, foramen ischiadicum majus; ***, future spina ischiadica; *red*, m. obturatorius internus; *blue*, m. levator ani; *green*, m. coccygeus

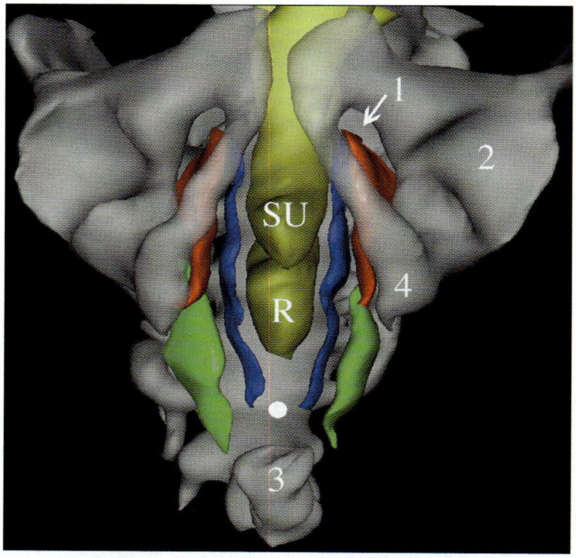

Fig. 31 Frontal view of the pelvic diaphragm. *R*, rectum; *SU*, sinus urogenitalis; *1*, foramen obturatum; *2*, acetabulum; *3*, os coccygis; *4*, ramus ossis ischii; *white dot*, future ligamentum anococcygeum; *red*, m. obturatorius internus; *blue*, m. levator ani; *green*, m. coccygeus

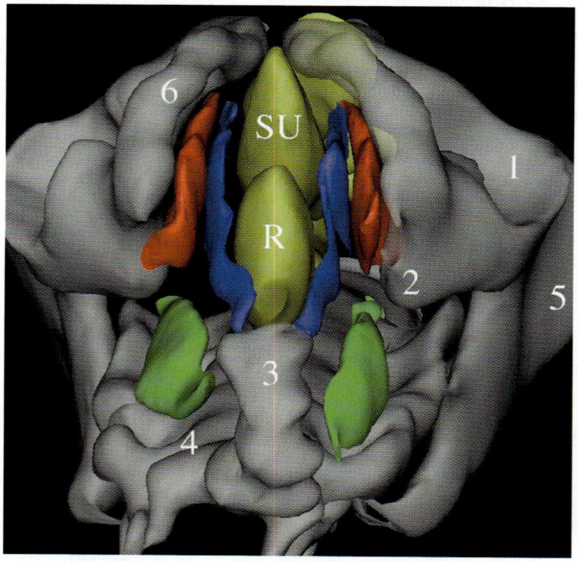

Fig. 32 View of the pelvic diaphragm from below. *R*, rectum; *SU*, sinus urogenitalis; *1*, acetabulum; *2*, ramus ossis ischii at level of developing spina ischiadica; *3*, os coccygis; *4*, os sacrum; *5*, os ilium (ala ossis ilii); *6*, ramus inferior ossis pubis; *red*, m. obturatorius internus; *blue*, m. levator ani; *green*, m. coccygeus

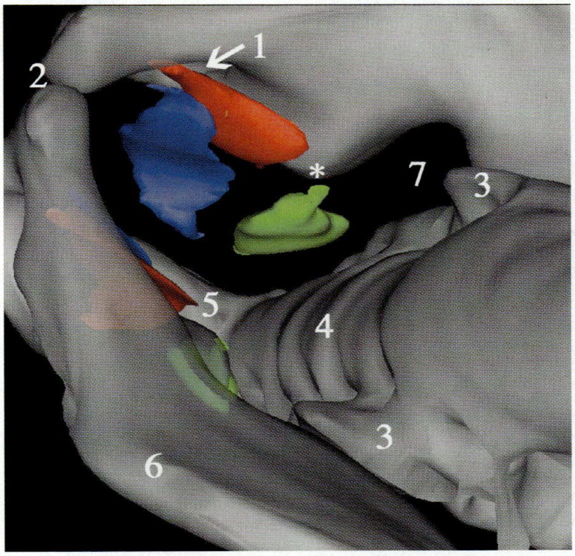

Fig. 38 View from above on the pelvic diaphragm. *1*, foramen obturatum; *2*, symphysis pubica; *3*, pars lateralis os sacrum; *4*, os sacrum; *5*, os coccygis; *6*, ala ossis ilii; *7*, foramen ischiadicum majus; *, future spina ischiadica; *red*, m. obturatorius internus; *blue*, m. levator ani; *green*, m. coccygeus

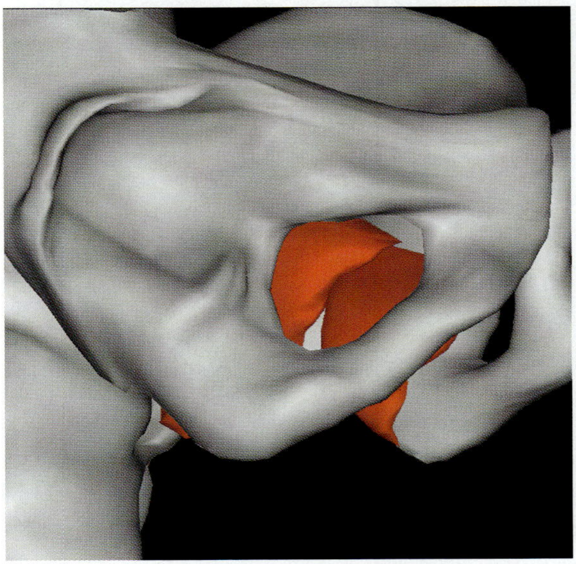

Fig. 39 View of the right foramen obturatum. The m. obturatorius internus is reconstructed in *red*. Note that the muscle does not cover the frontal half and cranial quarter of the foramen obturatum

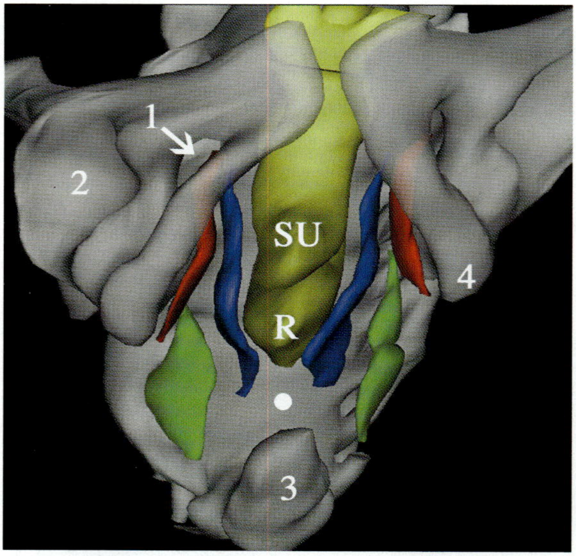

Fig. 40 Frontal view of the pelvic diaphragm. *R*, rectum; *SU*, sinus urogenitalis; *1*, foramen obturatum; *2*, acetabulum; *3*, os coccygis; *4*, ramus ossis ischii; *white dot*, future ligamentum anococcygeum; *red*, m. obturatorius internus; *blue*, m. levator ani; *green*, m. coccygeus. Note the absence of the curving between the sinus urogenitalis and the rectum (the future centrum tendineum)

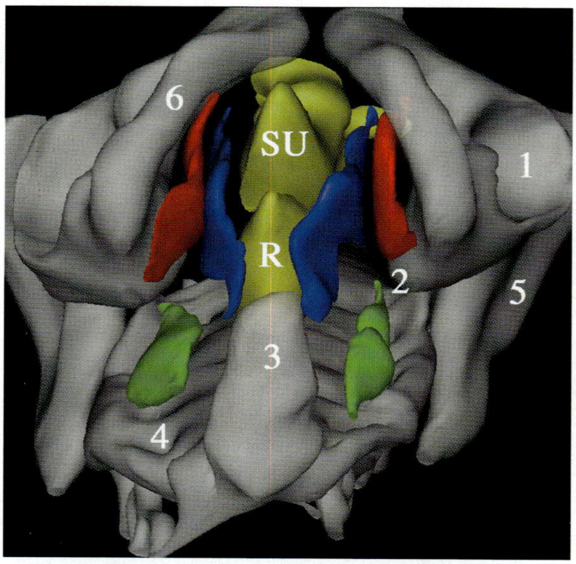

Fig. 41 View of the pelvic diaphragm from below. *R*, rectum; *SU*, sinus urogenitalis; *1*, acetabulum; *2*, ramus ossis ischii at level of developing spina ischiadica; *3*, os coccygis; *4*, os sacrum; *5*, os ilium (ala ossis ilii); *6*, ramus inferior ossis pubis; *red*, m. obturatorius internus; *blue*, m. levator ani; *green*, m. coccygeus

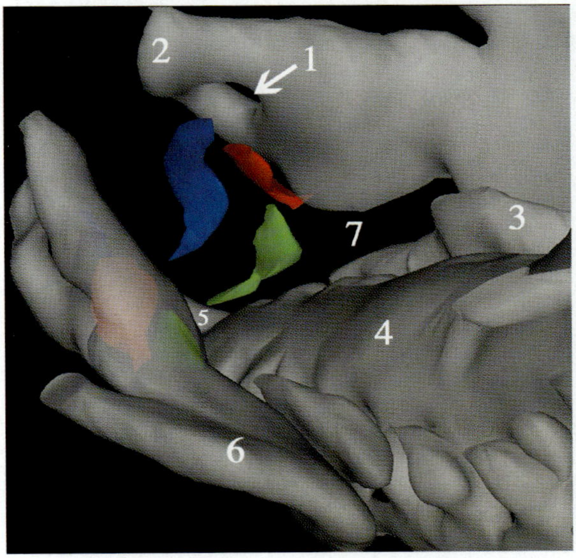

Fig. 47 View from above on the pelvic diaphragm. *1*, foramen obturatum; *2*, corpus ossis pubis; *3*, pars lateralis os sacrum; *4*, os sacrum; *5*, os coccygis; *6*, ala ossis ilii; *red*, m. obturatorius internus; *blue*, m. levator ani; *green*, m. coccygeus

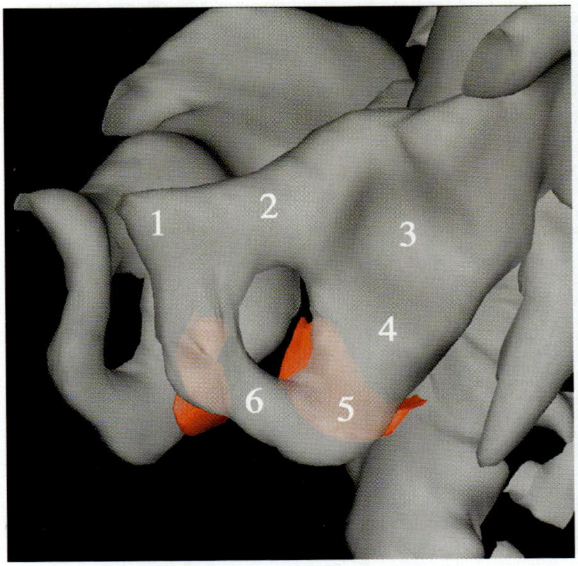

Fig. 48 Close view of the left foramen obturatum. *1*, corpus ossis pubis; *2*, ramus superior ossis pubis; *3*, acetabulum; *4*, corpus ossis ischii; *5*, ramus ossis ischii; *6*, ramus inferior ossis pubis; *red*, m. obturatorius internus

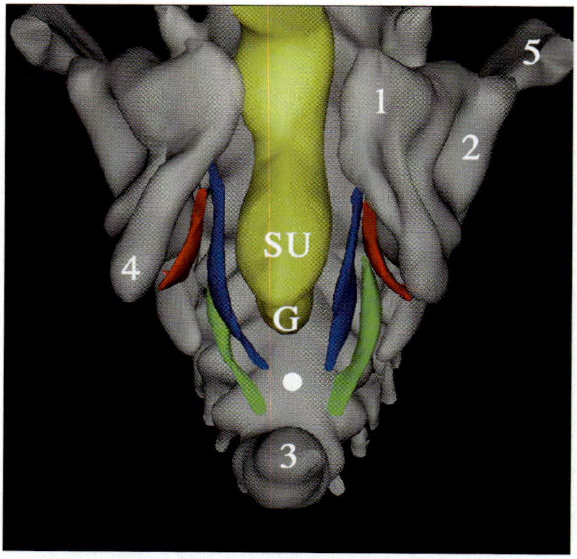

Fig. 49 Frontal view of the pelvic diaphragm. *G*, hindgut; *SU*, sinus urogenitalis; *1*, corpus ossis pubis; *2*, acetabulum; *3*, os coccygis; *4*, ramus ossis ischii; *5*, ala ossis ilii; *white dot*, future ligamentum anococcygeum; *red*, m. obturatorius internus; *blue*, m. levator ani; *green*, m. coccygeus

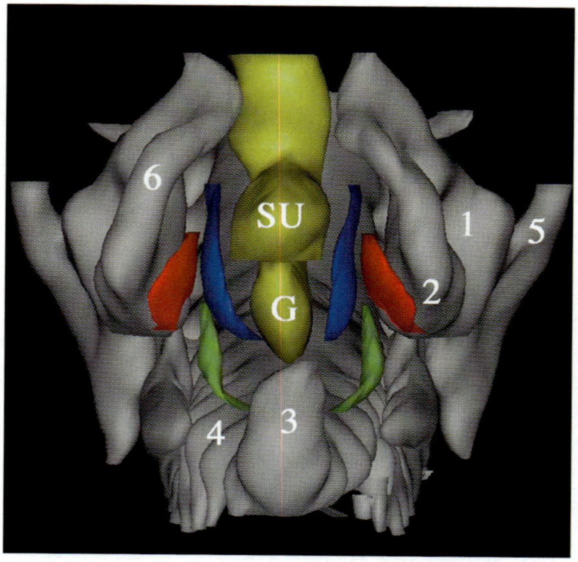

Fig. 50 View of the pelvic diaphragm from below. *G*, hindgut; *SU*, sinus urogenitalis; *1*, acetabulum; *2*, ramus ossis ischii at level of developing spina ischiadica; *3*, os coccygis; *4*, os sacrum; *5*, os ilium (ala ossis ilii); *6*, ramus inferior ossis pubis; *red*, m. obturatorius internus; *blue*, m. levator ani; *green*, m. coccygeus

References

Abitbol MM (1988) Evolution of the ischial spine and of the pelvic floor in the Hominoidea. Am J Phys Antropol 75:53–67

Abitbol MM (1989) Sacral curvature and supine posture. Am J Phys Anthropol 80:379–389

Abrams P, Cardozo L, Khoury S, Wein A (eds) (2005) Incontinence—3rd international consultation on incontinence, 3rd edn. Vol. 1 & 2. Health Publications, Paris

Barber MD, Bremer RE, Thor KB, Dolber PC, Kuehl TJ, Coates KW (2002) Innervation of the female levator ani muscles. Am J Obstet Gynecol 187:64–71

Barnes AR (1921) The pelvic fascia. Anat Rec 22:37–55

Barbaric ZL, Marumoto AK, Raz S (2001) Magnetic resonance imaging of the perineum and pelvic floor. Top Magn Reson Imaging 12:83–92

Bitoh Y, Shimotake T, Sasaki Y, Iwai N (2002) Development of the pelvic floor muscles of murine embryos with anorectal malformations. J Pediatr Surg 37:224–227

Blok S de (1982a) The connective tissue of the adult female pelvic region. Acta Morphol Neerl-Scand 20:65–92

Blok S de (1982b) Spatial architecture of musculo-fibrous tissue in the female pelvic region. Thesis: Amsterdam University

Bogduk N (1996) Issues in anatomy: the external anal sphincter revisited. Aust N Z J Surg 66:626–629

Bryce TH (1923) In: Bryce TH (ed) Quain's elements of anatomy, Vol 4, Pt 2, 1st edn. Longmans, Green, London, pp 205–218

Caro R De (1998) Morphometric analysis of the fibroadipose tissue of the female pelvis. J Urol 160:707–713

Cotran RS (1999) Genetic disorders. In: Cotran RS, Kumar V, Collins T (eds) Robbins pathologic basis of disease, 6th edn. W.B. Saunders, Philadelphia, pp 173–174

Chandebois R, Faber J (1983) Automation in animal development. A new theory derived from the concept of cell sociology. Monograph Dev Biol 16:1–202

Davies JW (1955) Man's assumption of the erect posture: its effect on the position of the pelvis. Am J Obstet Gynaecol 97:1012–1020

DeLancey J (1994) Functional anatomy of the pelvic floor and urinary continence mechanism. In: Schüssler B, Baessler K, Burgio KL et al (eds) Pelvic floor re-education. Principles and practice. Springer Berlin Heidelberg New York, pp 9–21

Delbet P (1907) Vessie, urèthre, prostate, verge, périnée. In: Poirier P, Charpy A (eds) Traité d'anatomie humaine. Vol 5, Pt 1. Masson, Paris, pp 74–240

Derry DE (1907) On the real nature of the so-called pelvic fascia. J Anat Physiol 42:95–106

Dickinson RL (1889) Studies of the levator ani muscle. Am J Obstet 22:897–917

Dorschner W, Stolzenburg JU, Neuhaus J (2001) Structure and function of the bladder neck. Adv Anat Embr Cell Biol; Vol 159:III–XII; 1–109

Elftman, HO (1932) The evolution of the pelvic floor of primates. Am J Anat 51:307–346

Fritsch H (1992) The connective tissue sheath of uterus and vagina in the human female fetus. Ann Anat 174:261–266

Fritsch H (1993) Development and organization of the pelvic connective tissue in the human fetus. Ann Anat 175:531–539

Fritsch H (1994) Topography and subdivision of the pelvic connective tissue in human fetuses and in the adult. Surg Radiol Anat 16:259–265

Fritsch H, Fröhlich B (1994) Development of the levator ani muscle in human fetuses. Early Hum Dev 37:15–25

Fritsch H, Hotzinger H (1995) Tomographical anatomy of the pelvis, visceral pelvic connective tissue and its compartments. Clin Anat 8:17–24

Fritsch H, Kühnel W (1992) Development and distribution of adipose tissue in the human pelvis. Early Hum Dev 28:79–88

Fucini C, Elbetti C, Messerini L (1999) Anatomic plane of separation between external anal sphincter and puborectalis muscle. Clinical implications. Dis Colon Rectum 42:374–379

Gardner ED, Gray DJ, O'Rahilly R (1986) Anatomy. A regional study of human structure, 5th edn. W.B. Saunders, New York, pp 499–501

Gegenbauer C (1903) Lehrbuch der Anatomie des Menschen, 7th edn., Vol 2. W. Engelman, Leipzig, pp 93–200

Gennaro M de, Mosiello G (2005) The pelvic floor—congenital anomalies from children to adults. Edizioni Goliardiche, Trieste

Gray H, Williams PL (1995) Gray's anatomy. The anatomical basis of medicine and surgery, 38th edn. Churchill & Livingston, Edinburg

Gribnau AAM, Geijsberts LGM (1981) Developmental stages in the Rhesus monkey (*Macaca mulatta*) Adv Anat Embr Cell Biol 68:1–84

Haderer JM, Pannu HK, Genadry R, Hutchins GM (2002) Controversies in female urethral anatomy and their significance for understanding urinary incontinence: observations and literature overview. Int Urogynecol J Pelvic Floor Dysfunct 13:236–252

Hafferl A (1969) Lehrbuch der topografischen Anatomie. Springer, Berlin Heidelberg New York

Hamilton WJ, Boyd JD, Mossman HW (1966) Human embryology: prenatal development of form and function. Heffer, Cambridge

Havenga K (1998) Total mesorectal excision in rectal cancer surgery. Thesis, Leiden University

Hjartardottir S, Nilsson J, Petersen C, Lingman G (1997) The female pelvic floor: a dome—not a basin. Acta Obstet Gyneacol Scand 76:567–571

Holl M (1894) Zur Homologie der Muskeln des Diaphragma pelvis. Anat Anz 10:395–400

Holl M (1897) Die Muskeln und Fascien des Beckenausganges. In: Von Bardeleben KH (ed) Handbuch der anatomie, Vol 7, Pt 4. Teubner Verlag, Jena

Jong de TPVM (2001) Surgical aspects of pediatric urinary incontinence. Thesis, Utrecht University

Klutke CG, Siegel CL (1995) Functional female pelvic anatomy. Urol Clin North Am 22:487–497

Knauer S (1914) Ursachen und Folgen des aufrechten Ganges des Menschen. Anat Hefte 22:138–141

Lawson JON (1974a) Pelvic anatomy (I) Pelvic floor muscles. Ann R Coll Surg Eng 54:244–252

Lawson JON (1974b) Pelvic anatomy (II) Anal canal and associated sphincters. Ann R Coll Surg Eng 54:288–300

Levi AC, Borghi F, Garavoglia M (1991) Development of the anal canal muscles. Dis Colon Rectum 34:262–266

Marani E (2002) The pelvis: another view. Twente University Press, Enschede

Marani E, Pijl ME, Kraan MC, Lycklama a Nijeholt GA, Videleer AC (1993) Interconnections of the upper ventral rami of the human sacral plexus: a reappraisal for dorsal rhizotomy in neurostimulation operations. Neurourol Urodyn 12:585–598

Meyer GH (1861) Lehrbuch der Anatomie des Menschen. Engelmann, Leipzig

Miller A (1996) In: Rimoin DL, Conner JM, Pyeritz RE (eds) Principle and practice of medical genetics, 3rd edn. Churchill Livingston, New York, pp 245–246

Mostwin LM (1991) Current concepts of female pelvic anatomy and physiology. Urol Clin North Am 18:175–195

Occelli B, Narducci F, Hautefeuille J (2001) Anatomic study of arcus tendineus fasciae pelvis. Eur J Obstet Gynecol Reprod Biol 97:213–219

Olsen AL, Rao SS (2001) Clinical neurophysiology and electrodiagnostic testing of the pelvic floor. Gastroenterol Clin North Am 30:33–54; v–vi

O'Rahilly R (1979) Early human development and the chief sources of information on staged human embryos. Eur J Obstet Gynaecol Reprod Biol 9:273–280

Piersol A (1930) Human anatomy, 9th edn. Lippincott, Philadelphia, pp 557–562

Pit MJ, de Ruiter MC, Lycklama A Nijeholt AA, Marani E, Zwartendijk J (2003) Anatomy of the arcus tendineus fasciae pelvis in females. Clin Anat 16:131–137

Popowsky J (1899) Zur Entwicklungsgeschichte der Dammuskulatur beim Menschen. Anat Hefte, Abt. 1 12:13–48

Power J (1948) Embryological development of the levator ani muscle. Am J Obstet Gynaecol 55:367–381

Putte SCJ van der (2005) The development of the perineum in the human. A comprehensive histological study with a special reference to the role of the stromal components. Adv Anat Embr Cell Biol 177:1–131

Quain J (1899) Elements of anatomy, Vol 2, Pt 2. Longman, Green London, pp 343–344

Shafik A (1975) A new concept of the anatomy of the anal sphincter mechanism and the physiology of defecation. II. Anatomy of the levator ani muscle with special reference to the puborectalis. Invest Urol 13:175–182

Thompson P (1899) The myology of the pelvic floor. A contribution to human and comparative anatomy. McCorquodale, London

Vesalius A (1543 and 1555) De Humani Corporis Fabrica, Libri Septem. Apud Basileae Oporinus

Waldeyer W (1899) Das Becken. Friedrich Cohen, Bonn, pp 209–211

Zacharin RF (1985) Pelvic floor anatomy and the surgery of pulsion enterocele. Springer, Berlin Heidelberg New York, pp 31–38

Subject Index